Freyja Frimannsdottir
PR & Communication Manager for IKEA Group
"This book addresses an important issue – collaborating between corporations and universities in pursuit of large-scale societal issues such as food waste and loss. It is great reading, and a good manual, for such collaboration. I highly recommend the book for anyone in a private or public organization seeking to tackle grand challenges, wicked problems and other issues of shared importance. Collaboration is key to emergent solutions."

Thomas Candeal
Project Manager – International Food Waste Coalition
"This programme is effective – students work with an organisation in interdisciplinary, international teams, exploring solutions to real sustainable business challenges. They learn what it means to create value in society. I can't think of a better approach to help these students face world challenges when they begin their careers."

Johanna Lilius
Manager, International Affairs, Hanken School of Economics
"This book speaks to the need of universities to become international in a creative and innovative way, offering a helpful and pragmatic guideline to developing leading strategies for international studies. It is recommended reading for those in charge of international curricula and university partnerships and great reading for all pedagogical innovators looking for new approaches for how to practically implement new pedagogies."

Lorenzo Cantoni
Pro-Rector for Education and Students' Experience, USI – Università della Svizzera italiana
"Two key issues make this book so interesting for higher education. First: addressing major societal issues, with an interdisciplinary approach. Second: doing so with international students' teams, making the best of communication technologies."

Dany Stauffacher
CEO & Founder S.Pellegrino Sapori Ticino
"This book speaks and educates to think outside the box. Young people, who are going be the ruling class of our future, are finally called to face real and tangible problems of the real world with the possibility to develop a critical awareness also in the world of food, too often underestimated despite its central position in the lives of all individuals."

THE SUSTAINABILITY GRAND CHALLENGE

How do universities tackle wicked sustainability challenges faced by society?

The Sustainability Grand Challenge is a toolkit for setting up and running an interdisciplinary master-level course in the context of real-world problems such as food waste and loss. The book offers a new pedagogical approach that we call 'wicked' because it is unorthodox, ambitious, and tackles complex problems that won't go away. The pedagogy is also international at the course level rather than the conventional exchange semester, enabling institutions to embed international approaches to their core teaching.

The Sustainability Grand Challenge speaks directly to academics who are looking for solutions that provide stimuli for research and teaching while giving students an innovative, international learning experience. The approach develops student understanding of the UN Sustainable Development Goals as broad-scale societal issues which are difficult, if not impossible, to 'solve'. An important outcome of this approach is the laboratory-style classroom that creates opportunities for faculty, students, and companies to co-create solutions that are immediately implementable. The resulting methodology is based on industry–university collaboration (such as IKEA and Nestlé). The methodology is of interest to corporate leaders pursuing sustainability goals and business transformation.

Achieving sustainability requires cross-boundary, cross-disciplinary, experimental approaches that allow for scalability. Wicked problems can only be tackled with wicked solution approaches.

Michael Gibbert is Professor of Sustainable Consumption and director of the World Challenge Program at Università della Svizzera italiana in Switzerland.

Liisa Välikangas is Professor of Leadership at the the Technical University of Denmark, DTU. She is also affiliated with Hanken School of Economics in Finland.

Marijane Luistro-Jonsson is Post-Doctoral Researcher at Stockholm School of Economics in Sweden, where she teaches and conducts interdisciplinary research on the behavioural dimension of sustainability.

THE SUSTAINABILITY GRAND CHALLENGE

A Wicked Learning Workbook

Edited by Michael Gibbert, Liisa Välikangas and Marijane Luistro-Jonsson

Routledge
Taylor & Francis Group

LONDON AND NEW YORK

First published 2021
by Routledge
2 Park Square, Milton Park, Abingdon, Oxon OX14 4RN

and by Routledge
52 Vanderbilt Avenue, New York, NY 10017

Routledge is an imprint of the Taylor & Francis Group, an informa business

British Library Cataloguing in Publication Data
A catalogue record for this book is available from the British Library

Library of Congress Cataloging-in-Publication Data
Names: Valikangas, Liisa, editor. | Gibbert, Michael, editor. | Luistro Jonsson, Marijane, 1971- editor.
Title: The sustainability grand challenge : a wicked learning workbook / edited by Michael Gibbert, Liisa Välikangas and Marijane Luistro-Jonsson.
Description: Milton Park, Abingdon, Oxon ; New York, NY : Routledge, 2021 | Includes bibliographical references and index.
Identifiers: LCCN 2020022944 (print) | LCCN 2020022945 (ebook) | ISBN 9780367197612 (hardback) | ISBN 9780367197629 (paperback) | ISBN 9780429243127 (ebook)
Subjects: LCSH: Sustainability–Study and teaching (Graduate) | Universities and colleges–Graduate work–Curricula.
Classification: LCC GE196 .S845 2021 (print) | LCC GE196 (ebook) | DDC 658.4/080711–dc23
LC record available at https://lccn.loc.gov/2020022944
LC ebook record available at https://lccn.loc.gov/2020022945

ISBN: 978-0-367-19761-2 (hbk)
ISBN: 978-0-367-19762-9 (pbk)
ISBN: 978-0-429-24312-7 (ebk)

Typeset in Bembo
by Taylor & Francis Books

To World Challenge Tacklers – all students (past, present and future) who aspire to be challenged and themselves challenge the world's persisting problems.

CONTENTS

ILLUSTRATIONS

Original drawings by Inês Costa.

Figures

Boxes

FOREWORD

Why it is worth investing in world challenges?

Stephanie Huber
Project Manager
Stiftung Mercator Schweiz

In the summer of 2019, the rainforest in the Amazon was in flames. In Iceland, the first glacier had to be declared dead. The *Special Report of the Intergovernmental Panel on Climate Change* (IPCC), which was published in the same year, reaffirms the worrying impact of our land use and food habits on the climate.

However, climate change is only one of many global challenges that need to be addressed to achieve the Sustainable Development Goals (SDGs) set out in the 2030 Agenda. Moreover, it is clear that many of the challenges can only be addressed with the help of a united force, that is to say, in various cooperative initiatives across disciplines, actors and countries. Such collaborations are essential, for example, to make global value chains more sustainable or to provide food security for the entire world population.

Accordingly, it is important that young people develop intercultural competencies and learn to work together across great distances in international, interdisciplinary teams. At the same time, it is equally important to find ways to deal with complex problems and the associated uncertainties (e.g. climate), and to come to common solutions under these conditions.

Mercator Foundation Switzerland therefore supports projects such as the 'World Challenges Program', which prepares young people for the sometimes challenging, international, and for the sake of our planet, urgently needed collaboration. Due to the climate impact of air travel, the foundation particularly welcomes the testing of virtual forms of cooperation and innovative e-learning models.

Magnus Mähring
Professor of Entrepreneurship and Digital Innovation
Chair of the SSE House of Innovation
Stockholm School of Economics
Stockholm
2 April 2020

It is a delight to be asked to write the foreword for this Wicked Learning Work-book. The book – and the continuing endeavour it is based on, the Tackling World Challenges interdisciplinary, multi-campus master course collaboration – addresses a number of crucial challenges and opportunities for higher education; I am thrilled to have the opportunity to reflect briefly on how they tie to some learning principles and practices that are dear to me.

My excitement, I must admit, also stems from having been there at the incep-tion: an informal, largely improvised, intimate academic gathering about everything and anything, organised by Liisa and Michael; celebrating a different book, and conducting conversations over meals and during walks through wooded hillsides overlooking Lago Maggiore. We discussed the future of the university in the light of digitally enabled learning, and its role in the light of the global challenges to ecological sustainability, as well as broader sustainability challenges. This was in 2015, and as I write this, I realise that 2015 is both a time to wistfully recall as saner, but also a time where most of our current challenges already existed: increasing rifts in the fabric of society, political instability and the mainstreaming of extremism, the threat of pandemics, war and migration, famine, pollution and the escalating overconsumption of available resources.

Clearly, these challenges place new demands on higher education, because they require new skills and capabilities from our current and future graduates, and from us as educators. And, as academics, we do have an unshakable belief in the importance of knowledge, and in the ability of our talented and passionate students to move mountains, particularly when given useful tools for the task.

Out of these conversations grew the idea for a quite unique, tripartite colla-boration, between Hanken School of Economics, Università Swizzera Italiana Lugano, and the Stockholm School of Economics, to create and collectively run a master-level course that would provide a range of skills, capabilities and experiences that would enable our students to effectuate meaningful and concrete change in relation to global challenges.

Reflecting on the many insights amongst the chapters in this book, and with some insights of my own in the work underlying the course and the book, there are a few things that resonate particularly with me, partly because they reflect learning principles and practices that I have experienced and found true. I will put them in my words, but they should be visible for the reader engaging with the texts in this volume.

First, in setting out on a journey like this, the right amount of crazy helps. The authors point out that 'wicked problems can only be tackled with wicked solutions', and maybe this is so, but not the least important is the liberating effect of generating and throwing out wild ideas and insisting that they are doable, maybe even normal. Having people involved who thrive on being a bit crazy helps ensure that solutions, designs, and learning processes become sufficiently 'wicked'.

Second, trusting the process is essential. This does not mean ignoring details, or quality considerations or contingency planning; it means trusting that an open-ended learning process that allows for unforeseen, real, things to happen is essential for harnessing the potential of a real-life project and problem-based learning experience, and that such a process can be shepherded and has self-repairing features. In contrast, putting the learning experience in a rigorous, preconceived structure – and the learners in a straightjacket of well-intended directiveness – is bound to result in an Ersatz experience, where students are likely to find themselves frustrated by the eerie disconnect between what is real and what is almost real, and consequently feeling distrusted and disempowered.

Third, it is important to immerse yourself to inspire others. This means letting go of using expertise as a protective shield or a means for enforcing hierarchy, and instead embarking on a learning journey together, nurturing your curiosity and passion and letting them shine through, and sharing freely what you know in order to achieve great things together. Like the previous points, this requires a bit of courage, and a firm check on your self-consciousness. In the end, what matters is what the student teams achieve and learn, not whether they thought you were particularly brilliant for 15 seconds, or even 15 minutes.

To conclude, I would like to congratulate the authors on their achievements related to Tackling World Challenges course. I am grateful that some ideas generated five years ago have come to fruition, and delighted that the collaboration has also resulted in this book, chronicling the collaborative journey. It is my hope that it will inspire many similar innovation initiatives across higher education institutions. This means that I also have reason to congratulate the reader, who might soon embark on a similar but different, crazy but fulfilling, wicked journey.

PREFACE

As we finish this manuscript, the world has turned upside-down from the global pandemic. All three partner universities have suspended face-to-face classes and are scrambling to provide alternatives so that students can learn remotely. The challenge for teachers to deliver their lessons using technology is real, and the timing of a Wicked Learning Workbook could not be better. When you read this, the world will still be in the pandemic winter or suffering from the frostbite of the global shutdown. Anyone who has suffered from frostbite knows that it never really disappears. Exposed to the same conditions, it starts hurting just as badly, again. Unless you are covered.

But who buys a winter coat on a hot summer's day? It seems odd to be ruminating about long-term strategies for anything non-medical while governments have everyone on lockdown. Yet, part of the struggle in this period of social distancing is solving the very problem that we address in this book. A classroom setup that focuses on the flexibility of the learning environment and the resilience of the individuals (because they are buttressed by a strong and non-local virtual team) works, even in a pandemic. The TWC class has kept calm and carries on. As ever more wicked problems and challenges arise, virtual pedagogy, wicked and otherwise are increasingly needed.

When did it all start?

Distance learning is nothing new – it started as correspondence courses in the mid-19th century relying on postal services carrying assignments between professor and student (Tait, 2003). The first degrees earned through distance learning were offered in 1858 from the University of London, referred to as the 'People's University' by Charles Dickens because it effectively democratised education, giving underprivileged students access to higher education. This programme boasts prominent alumni including Nelson Mandela, who studied while he was in prison (Bell, 2010). Since then, remote learning has advanced steadily, moving along with technology, first the television, and then, most importantly, the Internet. Digital evolution went from online education as another way of conducting a correspondence course, to enhancing the classroom experience. Well, any kind of education really. So where will it end?

How will this pandemic affect higher education?

Now, remote learning is jumping ahead of technology, due to the needs of students of all ages whose schools and institutions have closed in an unprecedented move to slow the spread of this 2020 pandemic, which scientists say will be the first of many in the near future. Teachers are speed-learning how to teach online, themselves becoming remote learners. Students everywhere are logging on and learning the pitfalls and pleasures of learning from home.

There is no doubt that this quick build-up of remote learning capacity will not be abandoned once schools re-open. This may prove to be the impetus that professionals pushing e-learning have been seeking to promote blended learning, even within traditional programmes.

Carol Switzer
Managing Director
World Challenges Programme
USI

References

Bell, S. (2010, 10 June). External students at the 'people's university'. Retrieved from https://www.bbc.com/news/10285568

Tait, A. (2003). Guest editorial: Reflections on student support in open and distance learning. *The International Review of Research in Open and Distributed Learning*, 4(1). doi:10.19173/irrodl.v4i1.134

ACKNOWLEDGEMENTS

We cheer the visionary leadership of the three participating universities in supporting an ambitious early idea with enthusiasm and providing space for its implementation. We give particularly warm thanks to Dean Maj-Britt Hedvall at Hanken. We appreciate the support of SSE's President Lars Strannegård, Senior Executive Vice President Lars Ågen, Vice President for Degree Programs Pär Åhlström and Director of House of Innovation Magnus Mähring. Special thanks go to Nina Volles, former Managing Director of Stakeholder Management & Fundraising at USI, for her critical role in obtaining funding and Mercator Foundation Switzerland for their contribution. Thanks also go to our administrations for fitting the novel course approach into current processes. To our IT departments for extending their role of IT 'support' and actually becoming part of the TWC team. To our corporate partners (IKEA, Nestlé, Sodexo) for fitting this into their busy schedules, and to International Food Waste Coalition for early partnering. To San Pellegrino and Sapori Ticino's Dany Stauffacher for being such a great sounding-board on gastronomic matters and for much appreciated oenological contributions to final presentations. To the Villa Carmine staff and above all to Gaetano Tresoldi for keeping a stiff upper lip when 40-plus World Challenge Tacklers invaded their kitchen to cook up waste-free meals (the freshly made chocolate ice cream will be remembered). To Carol Switzer for excellent project management during this book's editorial process. And finally, thank you to all our fabulous teaching staff, including doctoral students and post-docs, who have worked with us with dedication and enthusiasm. Now is no time to stop.

ABBREVIATIONS

ECTS	European Credit Transfer and Accumulation System
Hanken	Hanken School of Economics
SSE	Stockholm School of Economics
TWC	World in the Making: Tackling World Challenges
USI	Università della Svizzera italiana

CONTRIBUTORS

Inês Costa is a Visual Designer and Illustrator. She works with UX Design and UX Research in the health tech industry and curates Herchive, a digital art history project about women and non-binary artists. With a bachelor's degree in Stage Design she holds several years of experience working as a designer both in-studio and as a freelancer.

Tatiana Egorova is a PhD candidate at the House of Innovation at Stockholm School of Economics. Her research centres on entrepreneurship, with a particular focus on immigrant and social entrepreneurs. Her current research project is focused on the comprehensive assessment of well-being of immigrant entrepreneurs. Tatiana is also involved in teaching several master-level courses on sustainability, leadership and innovation. She is a former serial entrepreneur and management consultant.

Gottfried Gemzell is a Digital Pedagogue for Stockholm School of Economics and has been a course developer, digital development manager, teacher, copywriter, creative director, brand consultant and tutor. He has worked with the biggest and the smallest brands in Sweden and everything in between. Gottfried holds a master's degree in Literature Science from Stockholm University and a Diploma in Copywriting, Communication and Marketing from the RMI-Berghs School of Communication.

Michael Gibbert is a Professor of Sustainable Consumption and Director of the World Challenge Program at USI. He earned his PhD from the University of St. Gallen, holds an MSc from the University of Stellenbosch (South Africa) and did his post-doctoral fellowship at Yale University. He also worked as a research assistant at INSEAD, Fontainebleau and as Associate Professor at Bocconi University

(2003–2010) in Milan and is the author or co-author, or editor of numerous journal articles and seven books (published by Wiley, Financial Times Press, Elgar, and Blackwell) on innovation, strategic marketing, organisational theory and research methodology.

Alexandre Grandjean is a PhD candidate at the University of Lausanne (Switzerland), specialising in the nexus between religion/spirituality and sustainability. With socio-anthropological perspectives, he investigates how biodynamic agriculture, an esoterically driven and practitioner-based agronomy, is 'popularised' by the Swiss wine-crafting population. He earned the Best Student Paper Award of for his contribution: 'Biodynamic wine-crafting and the "spiritualization" of agroecology: Case study from Switzerland' in 2019 from the International Society for the Study of Religion, Nature and Culture.

Sofia John is a post-doctoral researcher and lecturer at the Hanken School of Economics, where she also earned her PhD. Her research interests include diversity, organisational well-being, and sustainability in an international context. Her work takes a micro-level perspective as she is particularly interested in the experiences and perceptions of the individual.

Marijane Luistro-Jonsson is a post-doctoral researcher at the Stockholm School of Economics, where she teaches and conducts interdisciplinary research on the behavioural dimension of sustainability. Among her research foci are cooperation, nudges and knowledge resistance. She is affiliated with SSE's House of Innovation (HOI), Center of Sustainable Markets (Misum), Stockholm China Economic Research Institute (SCERI) and the Center for Social Sustainability at Karolinska Institute. She earned her PhD from SSE, and master's degrees in Development Studies from University of Auckland (New Zealand), Business Administration from Ateneo de Manila University (Philippines) and Environmental Management from Stockholm University (Sweden).

Monika Maślikowska is a PhD student at the Institute of Marketing and Communication Management (IMCA) at USI on a Swiss Government Excellence Scholarship and is committed to teaching, research and academic service. She holds a master's in Management, a bachelor's in Business Management and Engineering (Architecture and Urban Planning) from Wroclaw University of Science and Technology in Poland, has studied in Australia, Italy and Spain, and worked for global organisations in Luxembourg.

David Mazursky (PhD, New York University) is the K-Mart Professor of Marketing (Emeritus) at the Hebrew University, Jerusalem. He has published widely on topics such as consumer information processing, judgment/decision-making, creativity in products and advertising, interdisciplinary creativity research in music, fashion and the arts.

Anna Nyberg is an Associate Professor at Stockholm University. She earned her PhD from the Stockholm School of Economics, where she also worked as an Assistant Professor where she was programme director for the BSc in Business and Economics at SSE and taught and developed courses at a variety of levels. Anna was responsible for the development and implementation of an integrated sustainability track during the first four semesters of the BSc curriculum at the SSE.

Sampo Sauri is a Media Producer for the Hanken School of Economics, works with IT and communications, produces video for teaching, and has been an editor at the Finnish national broadcaster Yle. Sampo holds a bachelor's degree in Television and Radio Broadcasting from Metropolia University of Applied Sciences in Helsinki.

Carol Switzer is a management consultant and has worked in systems planning, sales and marketing, market research, curriculum support and project management across a wide range of business sectors including academia, health care, entertainment, and social development in the US, the UK, Italy and Switzerland. Carol holds an MBA from the University of California, Los Angeles and a bachelor's in Mathematics at the University of San Diego.

Liisa Välikangas is a Professor of Leadership at the the Technical University of Denmark, DTU. She is also affiliated with Hanken School of Economics in Helsinki, Finland and EMLyon Business School in Lyon, France. Previously she has been at Stanford University, London Business School and Keio University in Japan. She worked for Strategos, a strategic management consulting firm, and SRI International in California's Silicon Valley. Her current research projects focus on the business and societal implications of digital technologies and large-scale innovation networks, including how to tackle very large problems, i.e. world challenges. Liisa is Senior Editor of *Management and Organization Review* and is known for her book publications in strategic renewal and organisational resilience.

Nina Volles is co-founder and Managing Director of Paeradigms, an NGO focused on achieving transformational outcomes in higher education. She has been designing and implementing academic programmes and projects in complex, multi-stakeholder environments, specialising in building strong partnerships and mobilising resources in support of institutional development. Before joining academia, she worked for over ten years in communications for multi-national corporations. Nina studied International Relations and obtained her doctorate (DBA) in Higher Education Management from the University of Bath in 2018.

PART I

Setting: Wicked, scarcity, waste, experimental, anthropocene, bubbly

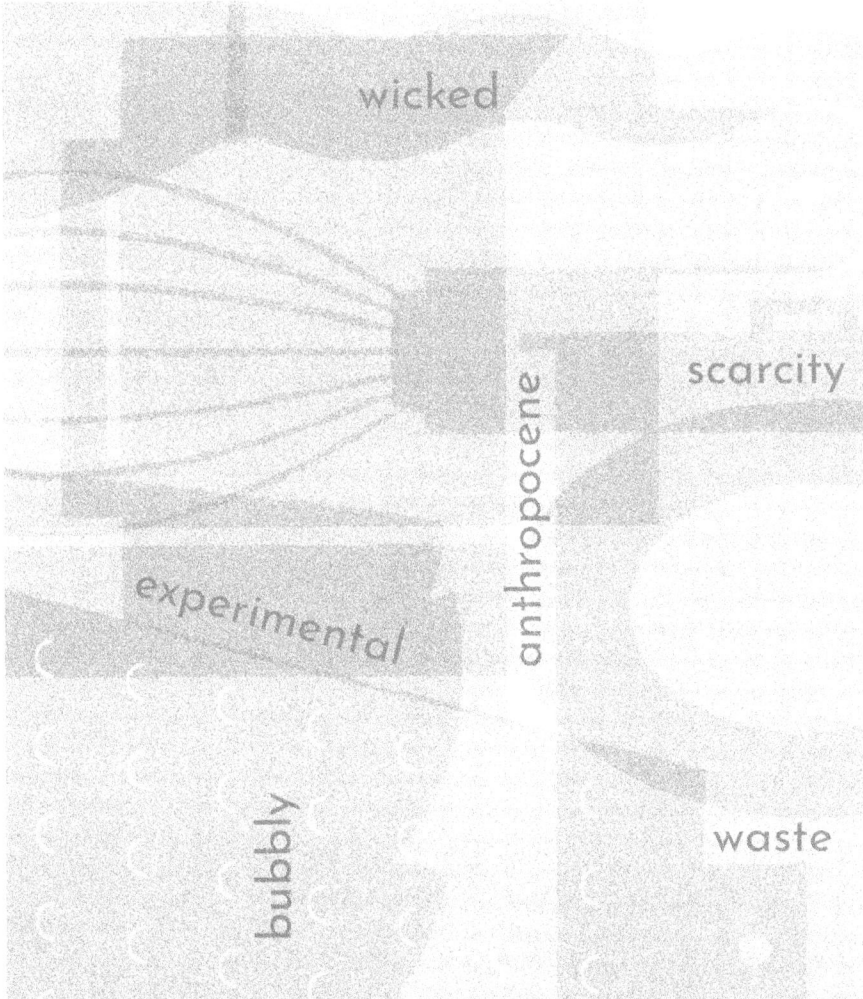

1

WICKED LEARNING

Why this course, and why now?

*Michael Gibbert, Marijane Luistro-Jonsson and
Liisa Välikangas*

How did it all start?

We started this course with a straightforward idea in mind: to give master's students the chance to work on 'wicked problems' – messy, ambiguous issues of global magnitude with no clear solution. Examples of wicked problems are migration, climate change, social inclusion, gender equality and food waste, which is what we discuss in some detail in this book. We hoped that students would rejoice in the opportunity to engage in such puzzles, especially if they are connected to sustainability and generally making the world a better place. As university teachers, we have a mandate for forming the next generations of decision makers in business and society. As such, we need to incorporate the changing environments and arising challenges, adapting syllabi and educational offers where possible in order to keep up with evolving demand from students and their future employers (Chapters 2–4 focus on the implications of the changing environment on tertiary education – including some ideas on how to implement new programmes and secure funding).

How can we tackle the world's biggest problems? We believe that there is value in confronting students with the inherent wickedness, or solution difficulty, of these issues from two perspectives (Chapters 11, 12 and 13 elaborate specifically on tools for tackling 'wickedness'). The first is pedagogical. Wicked problems require 'Wicked Learning'. In this Workbook we describe an indirect approach to problem solving where students learn to 'tackle' a problem, addressing the issue indirectly from a broader perspective such as the natural resources or animal rights, learning about it through engagement while involving different stakeholders including expert organisations, corporations, young people or teachers.

We decided to focus on a wicked problem which everyone (including students) will have been confronted with: food waste. Experimenting with 'starter solutions' nick away at the problem without trying to take it on all at once. For example, when approaching the problem of food waste in restaurants, one team of students looked at attitudes and behaviours of customers. What do we label as wasted food? Other teams designed an attractive recipe book for leftover food and an Instagram campaign for raising awareness of the amount of food that gets thrown out. They were surprised and delighted by the interest they received from thousands of followers when they launched. Part of the learning is to experience the difficulty in addressing these wicked problems, while increasing the understanding of the urgency and interdependency of the issues for our society. Most often, we need to find indirect solutions since indirect causes and impacts are part and parcel of the problem. A case in point: it is well documented that one-third of the food produced for human consumption (1.3 billion tons per year) is lost or wasted, using immense amounts of water unnecessarily and contributing significantly to global warming (FAO, Global Food Losses and Food Waste – Extent, Causes and Prevention, Rome, 2011 http://www.fao.org/3/a-i2697e.pdf).

The second objective for Wicked Learning is substantial, i.e. the topic of sustainability as the main common denominator underlying wicked problems. Students, upon graduating, will be confronted with demands of living more sustainably in all spheres of their professional as well as private lives. Sustainability – with wicked, or innovative approaches – will mark the next decade or so, the period where current master's students make their careers and raise their families. Their strategic decisions need to include innovative sustainability as a core criterion. So, the students – and their teachers – had better be ready.

Engaging in Wicked Learning, as we see it, requires crossing boundaries – disciplinary as well as national and cultural boundaries. As such, a key ingredient in our programme is to make students work in interdisciplinary and international teams (Chapters 7 and 10 provide essential guidance in this regard). To achieve this, we take students out of their comfort zones (away from their universities, core subjects and fellow students) and put them in cross-university teams (see Chapters 7, 8 and 9 for takes on topics such as virtual etiquette, keeping in touch from a distance, IT and other guidance on running such teams). In our case, this means that each team has students from all three partner universities. Since the universities are based in different countries, much of this collaboration must occur virtually as is the case increasingly in most global organisations. The course is open to students from all faculties, thus, these cross-university teams are also cross-disciplinary and students connect virtually with their teams during the semester while experimenting on a practical solution idea with the partner organisation or company. Team building is critical, so before students start collaborating in cyberspace, they meet in real space – each of the three participating universities takes turns hosting a kick-off event.

This virtual, cross-institutional, international and interdisciplinary collaboration has several implications. It mirrors closely what students will have to do in real life once they leave university. Indeed, one of our company partners noted when seeing our way of working: 'Oh, that is precisely how we work too!' A geographically dispersed team, with students majoring in disciplines ranging from finance to communications, allows students to experience a realistic work environment that they will likely encounter after graduation, for which conventional pedagogy leaves them utterly unprepared. It is ironic: we often preach the virtues of interdisciplinarity, the importance of working in international teams and the challenges involved in virtual collaboration, yet we rarely find hands-on courses which actually allow students to practise these skills.

To provide a 'realistic' environment where future skills can be fostered, we opted to collaborate with partner organisations (companies as well as NGOs, and Chapter 15 gives a candid account of some of the challenges involved). Together we decide on a particular wicked problem for a given course. The idea is to expose students to a hands-on experience where the partner organisation has an actual need to tackle an important sustainability issue that affects their business. In the case of the project we undertook with IKEA, for instance, the objective was to measure and reduce food waste in their store restaurants. The company had already started to work with production waste (i.e. what is wasted in the kitchens of store restaurants) and were beginning to look into plate waste (i.e. what guests leave on their plates after they are sated). Students helped the company by starting this process: collecting data, interviewing customers on site, observing behaviour around the food counters, and experimenting on different solution approaches.

Wicked Learning, therefore, introduces a new type of cross-border student experience which goes beyond traditional exchange and networking programmes (and Chapter 6 provides some insights into how to achieve critical buy-in at your own institution). The typical international exchange experience is based on the idea that students physically leave their 'home' universities to spend a period (usually a semester, sometimes a year) abroad, at some partner university. Wicked Learning, by contrast, integrates the international experience on the level and contents of the individual course rather than relegating it to several courses, usually electives, all taken at the partner university. This is no easy feat, especially when working with a multinational partner company. In the IKEA case, this meant that student teams had to figure out a way to collaborate with local IKEA stores, since students in each team came from three different countries and hence had to collect data in three different local IKEA restaurants, in three different languages, coordinating their own efforts with those of the other teams who also needed local presence. IKEA was an excellent partner in supporting the students, connecting them with local contacts, and attending virtual meetings with the cross-institutional student teams.

Thus, this new internationalisation model allows the university to achieve a qualitative, rather than quantitative leap when it comes to internationalisation strategy. The new metrics for measuring internationalisation no longer count how many students go on exchange programmes or how many faculty members carry a foreign passport, but how many courses students can take that are cross-university. Wicked Learning makes being international a core competency. To summarise:

1. The programme updates the conventional models of internationalisation based on the number of international students or student mobility and offers

the international experience as an elective course within an already existing master-level programme.

2. Its pedagogy reflects the work environment that professionals are required to operate in today; international experience is extended into interdisciplinary teams (different faculties) working cross-institutionally (three partner universities) – both face-to-face and virtually – using collaborative technologies. And field work with a partner organisation is coupled with academics, a scholarly study of the issue-at-hand and its potential solution paths, crossing the final boundary between society and university.

3. Wicked Learning develops well-rounded future leaders by investigating economic challenges within the context of broader 'World Challenges' – developing students' sensibility to the societal impact of their decisions (e.g. in the context of environmental degradation including climate change, politically motivated or religious violence, nutrition, poverty and population growth).

We believe that Wicked Learning provides the broad educational emphasis coupled with soft skills that result in well-rounded graduates. Students must be ambitious yet humble in addressing these issues critical to our time and well-being. Wickedness demands patience, smarts and determination. At the same time, it avoids easy, superficial, solutions that pretend to address the issue – in vain. Giving up is no option either; thus, the course seeks to maintain a generative attitude in our ability to cross open and closed borders and sometimes divisive boundaries to tackle these pressing challenges of our time. Being wicked also requires an attitude of persistence as the challenges cannot be solved overnight. Humour helps here too, wicked or otherwise.

Wicked Learning also equips students, individually and collectively, with skills that are conspicuously absent, and in our view crucially in demand, from most higher education curricula (and we are not afraid to share student voices and feedback in Chapter 14). We believe that the attitudes, strategies and skills students develop in their early to mid-20s are formative for the rest of their lives. As the famed MIT economist Paul Samuelson is claimed to have said referring to teaching young students: 'Get them early.' It is our hope that Wicked Learning will produce responsible decision makers who already have experience working in virtual, international teams, are able to think holistically by considering the broader context in which business decisions are made, and consider the impact of these decisions on key stakeholders and issues. Importantly the students gain the experience of getting engaged in changing the situation, one experimental step and one local or global partner at a time.

That is a wicked mission worth working on.

The (wicked) way forward

At the evening of the kick-off meeting of the 2018 course, we got together with the students over supper and biodynamic wine (with the accompanying story of sustainable wine production; see Chapter 5 for an unusual take on the subject), and the idea of sharing our experiences was born. Hence this book. Our approach to writing this Workbook is to give everyone who made it happen (students, industry partners, instructors, IT specialists, university leaders, as well as accreditation officials) a voice in framing their part of the experience. We believe Wicked Learning requires everyone involved and these are their stories and own Wicked Learning recipes from which others may be guided.

Welcome to the World of the Wicked!

Our approach

When we first had the idea of providing others with some kind of manual, some suggestions of how to do things (and how to not do things), in a nutshell of how to run such a course, we wondered which format such a 'manual' should take. We opted for an 'edited' book. The reason was simple. There are so many stakeholders necessary to make this course fly that we felt it was wrong to speak 'for' them. The edited, or curated, book format here is ideal as individual chapters are authored by those most knowledgeable about a certain topic. And there are many contributors from many different backgrounds, including

- the university leadership who generously provided the space (curriculum-wise) for this course in the first place;
- a representative of the funding organisation who provided part of the resources to make this fly;
- university admin personnel who engaged in internal marketing, making sure the course was on the agenda of the relevant master's directors, deans' offices and registrar office;
- IT professionals from our three universities who provided the infrastructure so that we could talk;
- PhD students and post docs who were most directly involved in the selection of the students, day-to-day running of the course, collecting student feedback and advising us about sustainable (and enjoyable) consumption habits more generally;
- and last-not-least even the editors themselves add a thought piece at the beginning and as a wrap-up of the book.

In selecting these contributions, we have encouraged each author to speak with her or his own voice (sometimes quite literally, e.g. in the case of interviews with rectors) and our objective here was to let that voice speak as authentically as possible (rather than shoehorning them in some kind of generic template). We have tried to maintain the authenticity as well as the love, sweat and tears that went into the programme as a way to illustrate a very plain key learning: it needs to be fun for everyone involved.

The result is a patchwork of contributions, which we stitch together as 'scenes' or parts. That is, we take a process perspective (from establishing the programme, pre-launch considerations, running the programme, and eventually some lessons learned in the form of food – and wine – for thought). We hope that these 'scenes' or parts will make it easiest for the reader to access the individual chapters which are most relevant to him or her. The academic course is called 'The World in the Making: Tackling World Challenges' and we refer to it throughout the book as TWC.

Part I: Setting: Wicked, scarcity, waste, experimental, anthropocene, bubbly

- Chapter 1 (Wicked Learning: Why this course, and why now?) brings together the three editors, who are faculty members from Hanken School of Economics, Finland; USI University of Lugano, Switzerland; Stockholm School of Economics, Sweden (Michael Gibbert, Liisa Välikangas and Marijane Luistro-Jonsson, respectively). Here, we take a step back to the very beginnings of this idea, and how we experienced, individually and jointly, the drawbacks of the current tertiary education system.
- In Chapter 2 (Global Challenges as necessary preparation for students), Anna Nyberg and Marijane Luistro-Jonsson share a fascinating account of extending the idea of Global Challenges to the bachelor level at SSE. This chapter is a must-read for anyone interested in the implementation issues a cross-disciplinary topic such as sustainability entails. As the authors put it, the first question we need to ask ourselves is whether to 'add sustainability content to the existing courses or to create new courses' (p. 4). Neither of the two seemed satisfactory to all stakeholders, as the first one crowds out within-course content and the second one eclipses entire courses. In the end, a third way as chosen, namely the 'Global Challenges Track', which consisted of four new mandatory courses, organised in three tracks: Knowing, Doing, Being, Expressing. Anna and Marijane not only describe the considerations which led into the complete overhaul of the Bachelor education at SSE but also relay the reactions of students, faculty and university partners.
- Chapter 3 (The proposal: How we got started) by Michael Gibbert, Carol Switzer and Nina Volles describes the process of obtaining outside support. Without the generous funding of Mercator, this project would have looked very different, and this chapter provides the funding agency's perspective. Basically, the funding allowed us to at least meet once (for the kick-off) and in some cases even twice (for the kick-off and the final summit where the participants presented their work in front of the partner organisations). Without it, the course, as we see it, is still doable (run as a completely virtual exercise) and we elaborate on this. But clearly, providing the students with a chance to meet their team members in real time and space helped make this a success.
- Chapter 4 (A new ambience of internationalisation, digitalisation and sustainability for students through pedagogy) by Carol Switzer and Marijane Luistro-

Jonsson gives space for the university leadership to express their thoughts. We went to our rectors and asked them where they see tertiary education going, what the role of sustainability in education is, how they see university–industry partnerships and how the course fits into the overall strategy of our three universities.

- Chapter 5 (Tackling World Challenges through renewed consumption habits) by Alexandre Grandjean describes the ambience and spirit of the course very well. Alexandre takes a fresh perspective on consumption by systematically comparing three different ways to consume wine ('natural' wine, that is). The three consumption forms are ingurgitating, swallowing and incorporating. This chapter provides some thought-provoking insights into the emerging natural wine culture but has wider ramifications in sustainable consumption – how can we be sustainable if we continue to simply swallow or even ingurgitate?

Part II: Ingredients: Blender, diversity, learning, international, encounters, broken

- Chapter 6 (The invitation: Timing is everything) by Carol Switzer is the first essential 'ingredient' in our recipe and shares some of our experiences in promoting this initiative internally. This is easier said than done since we are talking about a university-level elective, which cuts across departmental boundaries. Especially in universities (such as USI), where individual departments/faculties make their own decisions about which courses to offer students, a course such as TWC requires a fair amount of salesmanship and internal marketing to make decision makers (master's directors) appreciate the course's full potential for their students.
- Chapter 7 (A learning blender: How virtual team set-up influences outcomes) was written by Monika Maślikowska during her PhD at USI. Monika was the course administrator across the three universities and in this chapter relays the fundamental processes of forming teams that cut across university and departmental boundaries. It is essential reading for anyone working on the shop floor of this virtual/blended experience. The old fashioned title 'teaching assistant' is misplaced in this context and the chapter explains how and why.
- Chapter 8 (Virtual etiquette: Encounters among digital natives) was written by our self-declared 'digital pedagogue' Gottfried Gemzell and is a real gem for anyone working with teams in education, be they virtual or real. It is not at all technical even though Gottfried works in SSE's IT department (mind you, he has a degree in Literature Science from Stockholm University and a Diploma in Copywriting, Communication and Marketing from the RMI-Berghs School of Communication).
- Chapter 9 (Where are we going and how do we get there?) by Sampo Sauri, Media Producer at Hanken School of Economics in Helsinki, Finland is written from a technical IT perspective and illustrates the challenges and opportunities in finding the right platform for communicating across

universities (and even time zones, in our case). We experimented initially with Skype for Business and Sampo shows how Microsoft Teams eventually convinced us as the tool of choice (and was always available when anyone struggled – with passwords or otherwise). Sampo set up and maintained the digital infrastructure for the course communication.

Part III: Method: Collaboration, tools, creativity, resilience, challenge, salt

- Chapter 10 (Collaborative pedagogy and overcoming differences) by Sofia John (with input from Liisa Välikangas) is written from the perspective of the teaching team and describes the practical challenges faced in executing the course across international and institutional borders on the 'shop floor' of everyday teaching. To begin with, every course needs a teaching team comprising professors, teaching assistants, administrative staff, as well as technical support staff. The big change for TWC was that it spanned institutional boundaries, rules and regulations (including IT systems!). With such a large teaching team and a course that was ever evolving as it progressed, a crucial element of running the course proved to be quick and timely communication among the staff members. Secondly, communication and coordination is at a premium among such a large teaching team with new corporate partners located in different countries and different time zones every year. The third insight was that the course requirements had to be adjusted to fit the rules and systems of the different universities, and this chapter describes how we tried to tackle this particular wicked problem.
- Chapter 11 (A toolkit for Tackling World Challenges: Approaches and methodologies for teaching sustainability) by Marijane Luistro-Jonsson and Anna Nyberg shares the experiences of two university faculty members in teaching sustainability, and how TWC differs from other sustainability courses. Marijane and Anna share their insights with regard to the methodology for tackling wicked problems, in particular confronting two different approaches (direct and indirect). This chapter will be appreciated by anyone who is supposed to teach a method for solving problems that are inherently unsolvable. How to you explain this to your students? Equifinality in tool usage is already difficult enough to teach, but how do you make your students get excited about equifinality in a context where there are not only different tools but also different outcomes and none of them fully satisfactory?
- Chapter 12 (Wicked pedagogy as creative bricolage) by Michael Gibbert, Monika Maślikowska and David Mazursky uses bricolage as a metaphor to show how resources which are already at hand (in our case at your university) can be 'cobbled together' to create something new and exciting. Michael and David are professors of marketing and consumer creativity and as such draw on parallels from innovations that are based on existing resources, including those that do no longer work. This chapter will be inspirational for university

professors who need to innovate their pedagogy on a shoe string (who doesn't?). It is a manifesto for bricolage in the context of sustainability (which is inherently about using what is already available).

- Chapter 13 (Problems you solve and problems you work on: Connecting to, and engaging with, society and corporations) by Liisa Välikangas provides three essential pillars for tackling grand challenges. Important to her thinking is the role of 'resilience'. The key problem in tackling grand challenges is the risk that one experiment producing (apparently) the best outcomes at a particular moment in time will be copied and scaled up mindlessly, without other experiments and learnings in the future (if it ain't broke don't fix it, right?). In this chapter, Liisa draws on decades of experience in working directly with industry partners, and generously shares these insights in the form of three concrete steps to achieve resilient 'solutions' to grand challenges in collaboration with industry partners.

Part IV: Feedback: Enjoying, sharing, dinner, partnering, scaling, entrepreneurial

- Chapter 14 (Sharing student voices) by Monika Maślikowska, does, as the title suggests provide centre stage for students to talk about their experiences. After every instalment of the course, we asked the students questions about their learnings both from a process perspective (how did they experience the virtual learning experience, how did they find a common language in cross-disciplinary teams, how to avoid free-riding), as well as from a content perspective for instance, how did they experience the methodology and its applications more widely (after all, there are wicked problems to be tackled beyond food waste). Beyond the questionnaire-based data, Monika provides rich insights from interviews on the overall perception of the students, which were extremely helpful to improve bits and pieces along the way.
- Chapter 15 (Creating shared value for companies in Tackling World Challenges) by Tatiana Egorova and Marijane Luistro-Jonsson deals with one of the most important issues: the shared value created by the course for industry partners. Anyone who has worked in higher education in direct collaboration with industry partners will have experienced that industry partners often do not fully exploit the potential for joint learning (at least from the perspective of academics, that is). The authors explore possible reasons including the extra burden imposed on the employees involved, who will not directly benefit from the value created. Another issue is that since students are often not experts, a company may perceive their potential contributions as limited. To rectify this, the authors use a theoretical lens of shared value, and review the potential avenues of value creation and then discuss how the course in general, and the projects developed by students in particular, contribute to the creation of shared value from the perspective of the industry partner. The direct and upfront communication about the benefits offered (and the needed

commitment) arising in a collaboration between the industry and universities is of vital importance.

- Finally, far from concluding in any way this journey, Chapter 16 (Wicked Learning: An evolving recipe) provides some guidance for others who may wish to embark on a similar journey. In this final chapter, Marijane, Carol, Michael and Liisa sketch and share the recipe that emerged and invite the reader to be part of shaping an evolving course recipe and take a place at the table of Wicked Learning.

2

GLOBAL CHALLENGES AS NECESSARY PREPARATION FOR STUDENTS

Anna Nyberg and Marijane Luistro-Jonsson

Background

The Stockholm School of Economics (SSE) is a private business school established in the early 1900s, the first business school founded in Sweden. Its founders included the leading Swedish business families and firms of the time. Today SSE remains in a unique position in Sweden, consistently ranked first among business schools in the Nordic countries and counts over a hundred leading national and international firms among its corporate partners. The BSc programme in Business and Economics – the present-day continuation of the original *Civilekonom* programme – continues to hold a position among the most popular university programmes in Sweden, and its alumni are highly sought after by businesses in Sweden and abroad. One might wonder what possible motive there might be to change such a winning concept. Still, that is exactly what happened. This chapter will explain the motivation behind the overhaul, its content and how it was carried out.

Remaining current means evolving

It is no less true in business education than in the business world that developing and evolving is necessary for staying current. Businesses and other organisations in society face changing environments and new challenges. As business research develops, new research areas emerge, forming further incentives for the education to adapt. If the SSE as a business school wants to continue to provide a valuable education to its students and to the business community, then it has to evolve with changing demands.

Sustainability and responsible management

Teaching sustainability and responsible management is a good example of change being driven by development both inside academia from the outside, represented by the business world as well as changes from the inside, in what the most talented students expect from an education. Inside the SSE, research in the general area of environment and sustainability has a history spanning several decades at the school, involving research both within economics and within different disciplines of business administration. Growing from developments in research, environment and sustainability gradually came to be reflected in teaching as well. Thus, even at the starting point of the transformation project that we will describe, the area of sustainability was already represented as modules in several courses within the BSc programme.

An inflection point

The slow and steady growth in importance of sustainability in the BSc programme and at the SSE as a whole, reached an inflection point symbolised by two important initiatives: (1) signing the UN Principles for Responsible Management Education (PRME) and (2) beginning the work on overhauling the BSc Business and Economics programme.

In 2013, the SSE's became one of the now over 650 signatories to the UN PRME. The universities and business schools that are signatories to the PRME work to

> raise the profile of sustainability in schools around the world, and to equip today's business students with the understanding and ability to deliver change tomorrow … to ensure they provide future leaders with the skills needed to balance economic and sustainability goals, while drawing attention to the Sustainable Development Goals (SDGs) and aligning academic institutions with the work of the UN Global Compact. (Overview, n.d.)

At that time, the school had also begun its work to find a new model for its BSc in Business and Economics. There was a growing feeling that the programme as a whole needed to provide its students with a higher degree of competence in the area of sustainability and responsible management. These ideas and aims were shared by groups within the student body as well as within faculty, and increasingly were also voiced by the students' future employers in business and society. Building on work such as the so-called Second Carnegie report (Colby et al., 2011), there was growing realisation that the old way of teaching in business schools needed a change, not only for the sake of the students and their future employers but in the interest of the public good as well. Otherwise, business school alumnae and alumni would not be able to contribute successfully to the endeavours of their future employers and to society. Furthermore, as social responsibility increasingly gains importance for young people, the school's programme would eventually lose some of its ability to attract the most talented students. Finally, given SSE's standing, nationally and regionally, and the influential roles currently held by its graduates, the likelihood is high that changes in programme content will gradually have substantial effects on society. In 2015, SSE received funding for introducing a sustainability track in the BSc programme. The external and specific funding – awarded by the Global Challenges Foundation – would allow the SSE to recreate the programme in a way that retained its strengths while adding new competences for the students. The following pages will detail thinking and work in creation this change.

How did we do it?

A team consisting of teachers with experience from programmes at all levels of the school's education – bachelor, master, executive – took on the task of forming a proposition for the restructuring of the programme. A basic premise of the work was that it would be about 'change within the existing', keeping the existing strengths of the programme while infusing it with new, stronger sustainability content.

One of the first questions put to the group was whether to add sustainability content to the existing courses or to create new courses. In aiming to integrate sustainability into the overall programme the most straightforward way would seem to be to present sustainability content within the context of the established courses in the programme. However, an important challenge in this approach is that, especially in a decentralised organisation where each department and course director is the expert of their topic, such an approach would mean having to put the sustainability material in direct competition with existing course content. It was not clear that the new material would come out ahead in such a competition with what each course director and teacher team of existing courses presumably consider the 'core' of their respective course. Thus, considering the implicit challenges to important stakeholders, the end result with such an approach is hard to predict, both in terms of how well the material would truly be integrated and in terms of the longevity of the effort.

Instead, the group decided to add sustainability content to the programme through the creation of a 'Global Challenges' track consisting of four new mandatory courses required alongside the existing core courses. Each of the four courses addresses sustainability from a different broad perspective, not tied to a specific department. Each of the new courses has the explicit aim to create links to the existing courses. Thus, the programme and its existing courses are infused with sustainability content through cooperation with the new courses rather than through internal redesign. While this design is more acceptable to many stakeholders, it is also a design that increases resource demands on the school, as well as on the workload for students. Thanks to a generous donation from the Global Challenges Foundation, 40 MSEK (4.5 MUSD) over 10 years, SSE was able to introduce the new track without having to reduce the content or resources of any of the existing courses.

The new track is guided by the Global Challenges Advisory Board, bringing together SSE President and faculty, with international scholars including Jeffrey Sachs

(Columbia University), Johan Rockström (SRC), Kathrine Richardson (Copenhagen University), Johanna Mair (Hertie School of Governance), Ann-Sophie Crepin (Beijer Institute) and Kate Raworth (universities of Oxford and Cambridge).

The small team that developed the new Global Challenges track had representatives of the intended teachers at its core. The team cooperated and interacted with students, faculty and other staff throughout the development process. Members of the teacher team represent different academic backgrounds within economics and business, and this is an important asset in developing and teaching the courses.

The Global Challenges track involves four courses running over the first two years in parallel to other courses offered by a multidisciplinary team of teachers. The naming of the four constituent courses in the Global Challenges track – Knowing, Doing, Being, Expressing – indicates both course content and the progression of student involvement and competence in the area, moving from understanding challenges to confronting them (see Figure 2.1). An important goal in developing the course has been to include a variety of pedagogical approaches and forms of knowledge throughout. In addition to traditional academic knowledge built through lectures, seminars and literature, the courses include aesthetic and emotional knowledge built through interaction with different forms of art such as film screenings, museum visits and discussion of works of art in the school buildings. A team of teachers recruited from different subjects within the programme (i.e. economics, management, marketing, entrepreneurship) works together on the GC track courses, typically half of them focusing on the first two and the other half focusing on the last two courses. Half of the teachers from the very first year went on to teach the second year, and the aim was that teachers should continue to alternate between the sets of courses, although the latter has not been realisable. One challenge is the introduction of new teachers to the track, not least due to the fact that the courses are quite different from traditional core courses.

The aim of the first course (Knowing) is to develop basic knowledge about Global Challenges. The second course (Doing) aims to move into an action-oriented view, i.e. how to address global challenges through different types of

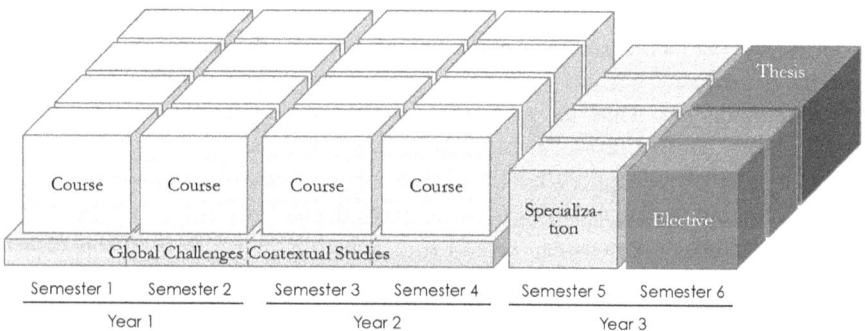

FIGURE 2.1 The Global Challenges track is distributed in existing courses over the first two years

organisations and social entrepreneurial efforts. In the third course (Being), when the students begin their second year at SSE, the focus is on the individual in relation to global challenges. The fourth course (Expressing) aims to bring the collection of courses together, by allowing the students to work on a project of their own choosing, where a specific global challenge is addressed through the means of a new social entrepreneurship venture.

Reactions to the introduction of the new track

Students' reactions

The reaction of students to the introduction of the new track varied among different stakeholders and over time. Overall, the reactions from the existing student body to the presentation of the programme were positive, although some voiced concern that the change would be too drastic or reduce the content in some of the traditional subject areas. When interacting with these students, it was very valuable to be able to point out that the new curses would not push out any of the existing core content. Reactions among the students actually taking the new track have been overall positive, as reflected in course evaluations, focus groups and individual reflection papers. A smaller number of students were extremely positive from the start, and a majority revealed a growing interest over time. Among development areas identified in the first years, many students would have liked an even stronger integration and a seamless approach to the Business and Economics programmes as a whole. Coming from a starting point where the core topics in the BSc programme were taught under the direction of independent departments, this vision by the students for a holistic programme represents a significant shift in thinking. The GC courses also shaped student expectations of a sustainability approach in other courses, thereby influencing content in these courses. Among prospective and incoming students, it is notable that the Global Challenges track is often brought up as having played a determining role in their choice of school. The new track has helped attract talented students that otherwise would not have considered applying to a business school such as the SSE.

Teachers' reactions

The early phases of teaching the track built a strong teaching team among the multidisciplinary group of teachers. The team was strengthened both from the joy of teaching and developing this entirely new set of courses, and from the challenges of the really very heavy workload involved in constructing and teaching courses unlike any other. The school supported the team by giving additional TAs and other support during the build-up phase. In this early phase, it was not entirely easy to recruit new teachers, even though those that did join the team were very enthusiastic. Now, the attitude has changed, and many more are interested in joining than are needed. The multidisciplinary composition of the teachers in the

CG track helps integrate different organisational silos within the programme as well as facilitates cooperation with parallel courses. Among colleagues in the school, the attitude was overall positive to the track and to forming cooperation across courses. Some scepticism regarding the effects on parallel courses existed, specifically some fear that the additional workload would have a crowding-out effect on students' attention in the existing courses. Over time, the new track has found its place in the total programme offering, strengthening the programme both through its own content and through cooperation with parallel courses.

External reactions

The new track has drawn a lot of attention outside the school, and as mentioned, helped attract students that would not otherwise have considered applying. Among other external parties, the many corporations and organisations that have generously supported the GC track by giving guest lectures, by welcoming study visits, and by hosting student projects, have been especially important to its success. The Global Challenges development, as well as the overall development of sustainability and training students to work with 'wicked problems', fit in to the strategic demand of the partner companies. To quote Lars Strannegård, the president of the SSE:

> We have 110 corporate partners. What they are basically all asking for – and that is why we are focusing more on sustainability issues – are questions connected to the SDGs in one way or another. Almost all of them want and need creative, analytical and emotionally mature individuals who are able to operate in an international context, who really understand these big problems. If you are able to come up with creative solutions, when it comes to food waste for instance, the students are interested to do it and the companies have many stakeholders including shareholders where reducing food waste will be an enormous economic upside if done in the right way (e.g. if through service rather than products). So, I see a great demand from the companies and our corporate partners for individuals who can solve problems like these. (Luistro-Jonsson, 2019)

Further developments

The overhaul of the Business and Economics programme and the introduction of the GC track were part of a bigger change programme at the SSE. The development of the TWC course is part of this development, and with this master-level course, the school now offers important sustainability content on both the bachelor and master levels. In the third year of TWC, we were very happy to welcome the first Global Challenges alumnus to the course.

In many ways, the TWC course supplements the Global Challenge programme. Although it is not as encompassing, it allows students to gather more knowledge, find out what others are doing, and express their own solutions to the challenges as they become the bearers of sustainability in their generation.

References

Colby, A., Ehrlich, T., Sullivan, W.M., Dolle, J.R. & Shulman, L.S. (2011). *Rethinking undergraduate business education: Liberal learning for the profession.* San Francisco: Jossey-Bass, The Carnegie Foundation for the Advancement of Teaching.

Luistro-Jonsson, M.J. (2019, 21 August). Personal interview of Lars Strannegård, President, Stockholm School of Economics.

Overview. (n.d.). Principles for responsible management education. Retrieved from www.unprme.org/about-prme/index.php

3

THE PROPOSAL

How we got started

Michael Gibbert, Carol Switzer and Nina Volles

Introduction

We first illustrated, in a nutshell, the main ideas behind the course. Thus, Tackling World Challenges is a second-year elective master's course in which 36 students from three European universities (USI, Stockholm School of Economics in Sweden and Hanken School of Economics in Finland) work in interdisciplinary and culturally diverse teams on a real-life 'wicked' problem as consultants to an interested organisation (2018–2020: food waste). We took care to nail down the main differences from the status quo when it comes to university internationalisation strategies. As such, we highlighted that the programme differs from existing internationalisation initiatives focusing on three specific goals:

1. Develop students' understanding of UN Sustainable Development Goals as broad-scale societal issues ('wicked problems'), which are difficult, if not impossible, to 'solve'. They require an extended reflection and application of problem solving techniques across disciplinary boundaries. The challenge we propose for the three-year pilot is 'food waste'.
2. Prepare students to work in globally dispersed teams. As companies expand geographically and technology is increasingly sophisticated, virtual teams are on the rise, and universities need to cater to this need (UKCES, 2014). Students remain based in their respective universities, and collaborate with their team across time, space and organisational boundaries. A number of programme modules and personal encounters serve to prepare students to work in virtual, multidisciplinary and diverse teams.
3. Pilot a didactic model that integrates international experience directly into the curriculum. This new internationalisation model allows the university to achieve a qualitative, rather than quantitative leap (not 'how many

international students are in your courses?' – but 'how international is the pedagogy of your courses?'). After a successful pilot, this model can be applied to other master's programmes at USI, and is scalable as a new element in other universities internationalisation strategy.

The Swiss context

We took care to situate the project in the local context (Switzerland). Thus, in the summer when we put in the application, IMD in Lausanne announced the findings of the 2016 edition of its World Competitiveness Yearbook (IMD, 2016), ranking Hong Kong as the world's most competitive economy; Switzerland was ranked second. The WEF's Global Competitiveness Report 2015 ranked Switzerland first overall, and fourth in higher education and training (World Economic Forum, 2015). We argued that while these rankings are positive, they represent a position established over the past decades. Switzerland will have to defend its position. Universities have a key role to play in terms of ensuring knowledge production through research and innovation and educating young adults for the needs of tomorrow (Smidt & Sursock, 2011). As such, we could illustrate to the funding agency how Swiss higher education institutions are faced with a number of challenges to defend their position, and how this project helps to tackle these challenges:

1. Limited access to international collaboration due to the initiative 'Against Mass Immigration'

    Switzerland is not part of the EU and as such faces a number of constraints when it comes to participating in international (EU funded) programmes. We used this fact to highlight the particular relevance of collaborating with other universities. In particular, Swiss higher education has been heavily penalised due to the adoption of the initiative "Against Mass Immigration" (February 9, 2014), which resulted in an immediate suspension from the majority of EU mobility and research programmes. The Swiss Government launched an interim solution called the "Swiss-European Mobility Programme", which funds student exchanges (outgoing and incoming). However, this solution does not compensate for Switzerland's exclusion from the EU's key actions to foster strategic partnerships, knowledge alliances and capacity building in higher education across Europe. (European Commission, 2016)

2. Integrating Sustainable Development Goals

    One of the key motivations behind the project was that universities are slow to adapt their curriculum to integrate sustainable development goals (SDG). Planet-wide problems do not recognise either national borders or the boundaries that have traditionally separated academic disciplines. We need to do a better job of developing young adults' understanding of complex, interconnected societal issues.

3.   Improving employability

The employability of recent graduates is a high priority at all universities. This concern transcends national boundaries and implementation priorities. However, several studies reveal that there is still much to be done to translate this priority into institutional practice (Crosier, Purser & Smidt, 2007; European Union, 2014; Fumasoli & Stensaker, 2013; Katsarova, 2015; Sursock, 2015). Today, students are ironically not well prepared for working in intercultural, multidisciplinary teams that are geographically dispersed. Recent studies suggest (Gilson, Maynard, Young & Vartiainen, 2015) that 66% of multinational organisations already utilise virtual teams. 80% believe this number will continue to grow. Universities need to integrate virtual collaboration skills into the curriculum. (Ferrazzi, 2014)

Our vision within this context

We proposed a qualitative departure from conventional models by integrating international experience into an existing master's programme, while fostering international collaboration via virtual encounters of students who remain based in their respective universities. On the substantive level, students work on a global issue of wide-ranging implications (UN Sustainable Development Goals), gaining a broad understanding of these complex 'wicked' problems and their impact on society. We took care to illustrate how the beneficiary university (only USI Lugano was eligible for funding given the policy of the funding body) is well suited to run such an initiative. As such, USI is a natural 'first mover' to test this new internationalisation model due to its experience as Switzerland's most international university, even though it has been particularly hit by the recent constraints in international collaboration by virtue of being small. Our approach is to take this challenge as an opportunity with the objective of testing a new model of internationalisation at USI for rollout to other universities in Switzerland.

Social change envisioned

Developing sensitivity for complex societal problems

The project is on 'World Challenges'. Our kick-off challenge, 'Food Waste', falls into the UN's Sustainable Development Goal #12, Responsible consumption and production. Students learn first-hand how businesses address complex world challenges and develop an awareness of the impact of their decisions on society.

Integrating training for working in interdisciplinary, multicultural virtual teams into the curriculum

This new teaching and learning model creates a new type of international student experience which goes beyond traditional exchange and networking programmes. The real-world setup optimally prepares students to work in interdisciplinary, virtual teams that are formed across the partner universities – a work environment that they will likely encounter after graduation, and for which conventional pedagogy leaves them utterly unprepared.

Creating a new internationalisation model

This new internationalisation model allows the university to achieve a qualitative, rather than a quantitative leap. After a successful pilot, this model can be applied to other master's programmes at USI, and is scalable as a new element in the internationalisation strategy for other universities in Switzerland and abroad.

What is the project about?

According to popular parlance, 'Necessity is the Mother of Invention'. In this spirit, TWC takes Switzerland's effective exclusion from mainstream EU led internationalisation programmes as a point of departure, and provides an alternative international experience that is fully integrated into the curriculum.

The partner universities are 'small' institutions and as such face particular challenges when it comes to positioning themselves against larger rivals, and when attempting to establish collaborative relationships due to their unequal bargaining power. Small size can be synonymous with elite scholarship and teaching excellence, which will benefit this project as it collaboratively explores big strategic opportunities. We strongly believe that in a world of Shanghai rankings and large universities, effectively making everyone the same, there is room for developing a unique strategy rallying small universities that share similar values, visions, and vantage points (Bleiklie, Enders & Lepori, 2012).

In particular, three opportunities are explored:

1. The programme updates the conventional models of internationalisation based on the number of international students or student mobility (European Parliament, 2015), and offers the international experience as an elective course within already existing master-level programmes.
2. Its pedagogy reflects the work environment that professionals are required to operate in today; international experience is extended into interdisciplinary teams (different faculties) working cross-institutionally (three partner universities) – both face-to-face and virtually – using collaborative technology.
3. It develops well-rounded future leaders by investigating economic challenges within the context of broader 'World Challenges' – developing students'

sensibility to the societal impact of their decisions (e.g. in the context of environmental degradation including climate change, politically motivated or religious violence, nutrition, poverty and population growth).

Individually, and together, the programme therefore equips students with skills that are conspicuously absent from most higher education curricula. We believe that the attitudes, strategies and skills students develop in their early to mid-20s are formative for the rest of their lives. The TWC will produce responsible decision makers who already have experience working in virtual, international teams, are able to think holistically by considering the broader context in which business decisions are being made, and consider the impact of these decisions on key stakeholders and societal issues.

To summarise, the TWC benefits can be grouped into two macro themes:

1. Pedagogical innovations:

    - Learning to balance economic and societal considerations.
    - Working in international, virtual teams across Europe.
    - Developing an integrated and interdisciplinary approach.
    - Independently managing contributions and accepting accountability to the virtual team members.

2. Educational innovations:

    - Switzerland is pioneering a new frontier of internationalisation in higher education in Europe, on the level of day-to-day pedagogy (as opposed to the status quo, where students go on exchange programmes without actual 'exchange' with their international peers during a course at their home institution).
    - Creating a new teaching and learning model that can be applied to other higher education institutions (which is highly scalable at limited marginal cost).
    - Investing in and extending clusters of excellent 'small' universities.

Programme structure and timeline

The TWC curriculum, repeated annually, includes lectures on how to tackle wicked problems and collaborate in virtual, interdisciplinary, multicultural teams; the teams are coached jointly by faculty throughout the semester-long course. The 36 students from three partner universities explore the challenge, access expert knowledge and engage in independent research. The student teams develop preliminary solutions and present their ideas in a Final Summit meeting. Students and faculty develop and learn methodologies to address complex, difficult issues building a strong team spirit. In particular, the course involves regular meetings in a variety of formats, from face-to-face to 'kitchen sink' work sessions among groups and the client, to virtual lectures, all of which are important milestones.

a   Kick-off session
    Location: Helsinki, Finland (alternating with Stockholm, Sweden)
    Length: 2.5 days
    Date: February of the second year of a master's programme
    Research on virtual teams strongly suggests that initial meetings drive positive
    team performance. As such, we start each spring with a 2.5 day face-to-face
    kick-off workshop to explain the programme, form the cross-institutional,
    multidisciplinary teams (two students from each university per team), and
    conduct team building activities. Specific focus will be on training students on
    how to approach wicked problems, collaborate in virtual teams, and work in
    an international, intercultural environment.
b   Meeting(s) with the International Food Waste Coalition and client organisation
    Location: varies according to client location and availability (Europe)
    Length: 0.5 day
    Date: March of second year of master's programme
    After several preparatory meetings with the client organisation (which is pro-
    posing the 'challenge'), the team will check in with the organisation in order
    to sharpen the focus on the issue, meet key stakeholders in person (to
    strengthen the relationship for further virtual work) and collect part of the
    data. Consideration will be given to scheduling this meeting face-to-face in
    diverse locations depending on client location with the aim of reducing travel
    where possible.
c   Final session
    Location: Lugano, Switzerland
    Length: 1.5 days
    Date: May of second year of master's programme
    During the final meeting, teams present their results to the client organisation
    as part of a one-day conference with students, faculty, and the client organi-
    sation, managers from the food industry as well as experts and opinion leaders
    in the field. Students and clients also evaluate the experience in terms of the
    objectives. A notable feature of the final session is that it has both a pedago-
    gical and an outreach objective. The outreach objective will be detailed in
    section 'Outreach activities'.

Commitment and consistency among university partners

All three university partners have committed to launch the TWC as an elective
course in the academic year 2017/2018 providing that start-up funding for the first
three years is obtained. As such, course credit had to be assigned to this elective.
What initially seemed like a no-brainer eventually developed (especially in the first
year of implementation) into a kind of wicked problem in its own right. The
conundrum involved two main issues: *inter-institutional heterogeneity* when it comes
to assigning credits (we used the 'European Credit Transfer and Accreditation
System', ECTS, where one ECTS accounts for 25–30 study hours). The second

was intra-institutional heterogeneity across individual master's programmes. Let's look at each in turn.

USI students form international, interdisciplinary virtual teams with student peers from the Hanken School of Economics in Finland and the Stockholm School of Economics in Sweden. USI students earn 6 ECTS credits for the course. At our partner universities, it emerged that elective courses such as this one would be 7.5 credits at SSE and 8 credits at Hanken (and later, 10 credits at Hanken). To resolve the issue of cross-institution heterogeneity when it came to ECTS assignment, we simply added additional work packages (e.g. additional questions in the final exam, or additional assignments in the interim reports). So that was fairly easy.

The issue of intra-institutional heterogeneity (across different master-level courses) was trickier. The course was originally planned as a six ECTS elective course that USI proposed to include in the second and final year of its master's programmes offered by all faculties of the university (architecture, communication, economics, and informatics, biomedicine). The problem was that the master's programme directors have the final say in whether a course can be acknowledged as an elective within their programmes. That makes a lot of sense, as programme directors need to 'protect' their students against additional work which may or may not be directly relevant to the goals of the overall curriculum. There was an underlying, institutional issue to this: students at USI were not allowed to select elective courses freely (the only constraint being timing, i.e. avoiding clashes with other courses); instead, the master's directors pre-selected a set of courses which were deemed sufficiently relevant (and non-overlapping with other courses). While some master's directors embraced the idea of sustainability, inter-university teams, etc., others found the course objectives only marginally suitable for their students and as such waived the option to include the course as a credit-bearing module in their programmes.

In the end, we resolved the issue in a pragmatic way, namely by offering the possibility to take the course as 'extracurricular' activity, i.e. not for course credit. This being said, the students needed to clarify with their master's directors the feasibility of obtaining course credit. This approach eventually satisfied everyone, as the primary benefit of the course for students was an extra item on their CVs, rather than simply another credit-bearing course. The additional benefit for us in the roll out phase was that it spared us the sales job of knocking on the door of every single master's director asking for permission to include the course in their formal curriculum. In a nutshell, IF students wanted credit, they had to ask for it themselves.

Reducing the ecological footprint and travel

We emphasised that the very goal of the TWC is to teach students (and their professors) how to effectively use the latest information and communication technology to minimise the need for (air) travel. As such, we are confident that the new tools we are teaching as part of the course will lead to greater awareness towards strategically keeping in person, real-time meetings to a minimum, i.e. at the beginning of the project, for team building purposes, consistent with the latest research on dispersed team collaboration.

We underscored the fact that food waste as such leaves a particularly strong ecological footprint (stemming from not only transportation but also water and other natural resources), and any attempt at reducing food waste will invariably reduce the negative ecological ramifications.

As to the logistics of the TWC, our aim has been to set up the programme in way to purposefully minimise the ecological footprint, both for the meetings among the students and professors, as well as for the meetings with partner organisations. Regarding student/professor meetings: since we are working with two Scandinavian partner universities (Hanken and Stockholm School of Economics), using the regular ferry boats between Helsinki (Hanken) and Stockholm will be the option of choice for moving students within Scandinavia. This has the added benefit of using the ca. 8–10-hour trip for additional project related activities including investigating ways to minimise food waste on ferry boats (the ferries within Scandinavia use the traditional buffet or smorgasbord, which has attracted some negative publicity in Scandinavia, precisely because it is so wasteful). This means that for meetings in Scandinavia (i.e. the kick-off and eventually interim meetings), only the Swiss delegation has to resort to air travel. Regarding meetings with partner organisations, we rely on virtual/digital resources. The virtual/digital aspects include student collaboration virtually using collaborative tools (such as Teams) and other interactive media (such as Skype). Some of the discovery work is done using tools that engage partner organisations in virtual interfaces (e.g. Adobe connect). Some of the teaching uses leading edge applications for asynchronous and synchronous interaction, discussion and presentations (e.g. establishing a permanent Teams profile as 'kitchen'). There will be digital tools that allow students to consult faculty in Internet-enabled virtual discussion rooms.

However, we also believe that the face-to-face meetings are key to successfully kick-off the project, ensure midway that everybody is on track, and close the programme presenting to and engaging with important stakeholders. To mitigate potential negative effects of lacking personal contact, each university participates in virtual class sessions from a digitally capable classroom, offering students from that location the possibility of meeting in person to connect virtually to the others.

Who is the target audience?

The project targets a number of main stakeholder group: students, university leadership, professors (including other universities), corporate managers, opinion leaders, NGOs, international organisations and policy makers.

Students

The project targets 12 USI students (annually) who are in the second year of one of the five joint master's programmes between the Faculties of Economics and

Communication Sciences. Participants are selected based on their potential contribution to the programme, grade point average, English proficiency, and the quality of their motivational letter and interview. Each year, the 'Tackling World Challenges' (TWC) class contains a maximum of 36 students, 12 from each partner university, and will be grouped into 6 interdisciplinary teams of 6 students.

Professors and university leadership

TWC professors at the three partner universities actively seek to raise awareness of the course to the academic leadership at their respective universities and generate interest among faculty colleagues. As we get more interest from USI students than available places, the initial objective has been to add another course to accommodate this 'spill-over' effect. Subsequently, we aim to expand the concept within USI, and provide support for the university leadership to propose the idea to other universities in Switzerland.

Corporate managers, opinion leaders, NGOs and international organisations

Involving corporate managers, NGOs, experts and international organisations (e.g. Global Compact) is an essential component to ensure knowledge transfer from 'outside' into the programme and the partner institutions and then back out to the 'outside world', which provides an arena for the potential application of student research.

Higher education policy makers

Ultimately, we aim to share our experience in developing and implementing the TWC with colleagues in Switzerland and Europe so that the maximum number of students may benefit from this innovative approach to extending their university experience. To do this, we planned to reach out to higher education policy makers, in particular the following two important forums:

a   Swissuniversities, which represents all Swiss universities, universities of applied sciences, and pedagogical colleges and promotes a common voice on educational issues in Switzerland. Apart from the rectors of the respective higher education institutions, there is also the 'Expertengruppe der Delegation Internationale Beziehungen' which consists of all Heads of International Relations.
b   ESKAS, the Swiss Federal Commission for Excellent Foreign Students, where the entire population of Swiss decision makers regarding internationalisation of Swiss universities can be reached. In addition, the president of this commission reports directly to the Swiss Parliament, thereby ensuring attention at the highest political level within Switzerland. Michael Gibbert was a member of ESKAS and as such could ensure that the message was spread.

How can we reach our target audience?

Students

The TWC is a second-year elective (i.e. part of the curriculum) of the master's programme and as such, is presented along with the general curriculum to students
during their first year. As part of the comprehensive communication plan, project
leaders collaboratively develop content for a separate electronic platform for which
university partner Hanken School of Economics (extensive experience in e-learning)
provides the expertise and infrastructure. Regular news via existing university outlets
further stimulates interest among future participants. As the programme is attractive
and places are limited, the demand repeatedly outnumbers the available places.

Professors and university leadership

USI and its partners

The programme involves three faculty members (one from each university partner)
who take the programme lead at their respective universities. The professors
actively seek additional faculty members to help with specific, disciplinary expertise
(for instance, quantifying food waste along the value chain from harvest to plate
waste may require expertise in finance, which the team is lacking). This 'buy-in' of
additional expertise occurs formally via faculty councils, i.e. the monthly meetings
where teaching and other administrative matters are discussed in the three universities. 'Buying-in' such expertise could lead to 'buy-in' from other colleagues,
hence extending the programme to other professors' leads to making the programme known in the three partner universities.

Other universities

As to sharing the WCP concept and experience with other universities, USI's
Head of International Relations has presented the programme at the Swissuniversities' 'Expertengruppe der Delegation Internationale Beziehungen' (consists of all
Heads of International Relations of Swiss universities). In order to ensure that the
information reaches the top leadership of Swiss higher education institutions, USI's
rector can share the TWC concept at the rectors' forum. We also used our own
networks to publicise the initiative. To ensure further outreach, Prof. Michael
Gibbert has reported on the progress of the programme during the yearly autumn
meetings of the ESKAS. Thus, every year in late November/early December, the
delegates from all Swiss universities (that is, professors who are responsible for
internationalisation at their universities plus the administrative staff including the
directors/secretary general of the international offices) are updated on the new
pedagogy, its implications for internationalisation in Switzerland, leading to a
reflection on a broader roll out among key decision makers.

Furthermore, the project group has submitted a concept paper for publication in a higher education journal about the new didactical approach of the TWC as well as a number of teaching methodology cases (including teaching notes) and a case book on food waste that function as knowledge dissemination tools.

Corporate managers, opinion leaders, NGOs and international organisations

The managers of the TWC partner companies are direct stakeholders in this project and are involved in and informed about the project outputs throughout the programme. Key opinion leaders and experts (often members an international organisation or an NGO – e.g. the international Food Waste Coalition) are consulted as part of the research. Moreover, the final meeting, held at the end of the programme, provides a perfect one-day platform to ensure knowledge transfer and exchange (one might even say a 'marketplace of ideas') between all major stakeholders, be they present or future. In fact, the final meeting also constitutes a 'friendraising' event where potential future donors can personally appreciate the outreach and impact of the programme.

Policy makers

Within Switzerland, there are two forums that effectively address higher education policy makers:

(a) Swissuniversities – which we can reach through the USI rector and our Head of International Relations and (b) ESKAS of which Prof. Michael Gibbert is a member (also see above).

How can we check progress and quality?

As part of the curriculum, faculty carefully elaborates learning objectives and outcomes, assessment methods and instructional strategy. Student assessments (tests, project reports, presentations) and evaluations as well as feedback from faculty and admin staff provide valuable feedback about learning outcomes, teaching methods, course strengths and areas for development. Adaptations for the next delivery of the programme are proposed, when appropriate. Examples of changes include modifications to course content, structure or to the assessment questions or methods. Ultimately, the project will be a success if, by the end of the pilot, other master-level programmes within USI have decided to add a similar elective course using the TWC as a model. Other master's programmes could, depending on their focus, tackle other UN Sustainability Development Goals (SDG), e.g. the master's in Health Management could address issues in the field of 'Good Health and well-being' (SDG goal 3) of the and the master's in architecture in 'Sustainable cities and communities' (SDG goal 11).

As part of the Swiss public higher education system and an accredited university, USI has a dedicated quality assurance department whose task it is to carefully monitor the institution's teaching and research activities. Each course goes through an evaluation process. The result is submitted to the dean and, if necessary, discussed in the faculty meeting. A further quality check will be the process of preparing findings for the rigorous publication process.

What could influence the project positively or negatively?

Opportunities

If the three universities succeed in building strong relationships with external stakeholders linked to 'Food Waste' (e.g. international organisations, interest groups, corporations, thought leaders), there could be a chance to build something relevant over several years that goes far beyond the student projects, involving internships, mentoring, research projects, etc.

The opportunity to enhance the student experience by integrating both relevant topics (food waste) with real-world working styles (intercultural virtual teams), will bolster the reputation of USI and the Swiss university system among prospective students.

Risks

The programme requires a minimum amount of funding to allow student teams to get a sense of 'being in this together', i.e. in particular the kick-off and final meetings. Team innovation literature consistently shows that on site meetings at these strategic moments are fundamental in ensuring successful team outcomes. At a very minimum, therefore, a kick-off and a concluding meeting should be held on an annual basis as proposed here to install and maintain momentum.

Since the TWC follows a different didactic model than the rest of the university, careful management of the pilot programme must be undertaken with adequate support for the faculty leaders in terms of technical platforms to facilitate a virtual working environment. Also, communicating the details of the programme, its inner workings, successes and challenges is essential to extending this model within USI and Switzerland.

Should the project continue after the expiry of the grant?

We proposed (and were awarded) seed money from Mercator Foundation Switzerland to cover pilot expenses incurred by Università della Svizzera italiana (USI) during the initial three years of operation of the programme. The requested funding will allow a total of 36 USI students (over a period of three years) to participate in this innovative initiative. During the pilot period, we will fine tune the project model, demonstrate success and secure funding from alternative sources. The ultimate goal is to propose TWC as a new student mobility and internationalisation model that can be adopted as a new teaching and learning model by other universities.

If the project continues, how will it be funded?

We do realise that the programme must become financially independent and would have three years to plan for this eventuality. As we want to keep the programme financially independent from the corporate sector, we believe we must explore several paths simultaneously in order to ensure long term financial viability:

1) Obtain public funding

> The first aspiration level of the project is national, rather than regional. Professor Gibbert, project leader at USI is also USI's delegate at ESKAS, the Swiss Federal Commission for Excellent International Students. If funding can be found to ensure successful kick-off and get-togethers in real-time and space, the project can serve as a best practice example of how CH can turn a challenge (the various funding constraints following the popular vote again mass immigration of 9 February 2014) into an opportunity, via crafting a qualitatively different model of internationalisation. The second aspiration level is international, by scaling up the model to become a separate funding format, similar to, e.g. the ERASMUS MUNDUS programmes fostering inter-university collaboration (albeit in the traditional sense, i.e. by funding entire new programmes, rather than funding courses within programmes). If the project is successful, we are confident that the first, national, aspiration level will at least be considered at the relevant authorities.

2) Create an event (e.g. the 'World Challenges Forum') to generate income to support the TWC going forward

> Here at USI, we have a track record of building and running conferences that bring together academia and the professional world. These conferences are an important part of our outreach strategy – beyond that, they can also be set up in a way (it usually takes a couple of years to get there) that allows us to reinvest the income into teaching activities. Some examples are: www.brand-management.usi.ch, http://efa2014.efa-online.org and www.emscom.usi.ch/sites/www.emscom.usi.ch/files/media/lac-leman-2011-invitation.pdf.
> USI's key values are 'international, interdisciplinary and innovative'. The TWC therefore fits perfectly well with USI's strategy.

Concluding thoughts and lessons learned

We are grateful to Mercator Foundation Switzerland for providing funds for the USI students to travel to the kick-off events in Scandinavia. Ideally, we would have ensured funding that covered all three universities. This unfortunately was beyond the remit of our grant. On the positive side, there was a pull-through effect in that the secured funding for the Swiss students afforded SSE and Hanken with a bargaining chip with their own university administration (which, however, has to be renegotiated every year).

While the funding agency generously provided resources for editorial support and a minimum of resources for (course) administration, it did not cover the faculty members' time spent in preparing, running and eventually teaching this course. Neither did the universities compensate the ECTS taught in some cases (via over-time teaching compensation or teaching discount). In other words, one of the professors effectively volunteered to teach this course on top of their usual teaching load without any compensation. We have all had (and are having) fun doing this, and this project is not about upping the monthly pay slip. On the other hand, getting some funds from somewhere to obtain a teaching buy-out is the way to go. So don't miss this item in your own funding application!

References

Bleiklie, I., Enders, J. & Lepori, B. (2012). Introduction: Transformation of universities in Europe. *Higher Education*, 65(1), 1–4. doi:10.1007/s10734-012-9577-5

Crosier, D., Purser, L. & Smidt, H. (2007). *Trends V: Universities shaping the European Higher Education Area. A European University Association Report*. Brussels. Retrieved from www.sowi-due.de/uploads/108.pdf

European Commission. (2016). Erasmus+ Programme guide. Brussels. doi:10.1109/CICEM.2013.6820130

European Parliament. (2015). *Internationalisation of higher education*. Brussels.

European Union. (2014). *The impact study: Effects of mobility on the skills and employability of students and the internationalisation of higher education institutions*. Luxembourg: Publications Office of the European Union. doi:10.2766/75468

Ferrazzi, K. (2014). Getting virtual teams right. *Harvard Business Review*, 92, 120–123.

Fumasoli, T. & Stensaker, B. (2013). Organizational studies in higher education: A reflection on historical themes and prospective trends. *Higher Education Policy*, 26(4), 479–496. doi:10.1057/hep.2013.25

Gilson, L.L., Maynard, M.T., Young, N.C.J. & Vartiainen, M. (2015). Virtual teams research: 10 years, 10 themes, and 10 opportunities. *Journal of Management*, 4(5), 1313–1337. doi:10.1177/0149206314559946

IMD. (2016). The 2016 IMD world competitiveness scoreboard. Lausanne: IMD. Retrieved from www.imd.org/uupload/imd.website/wcc/scoreboard.pdf

Katsarova, I. (2015). Higher education in the EU: Approaches, issues and trends. doi:10.2861/73364

Smidt, H. & Sursock, A. (2011). *Engaging in lifelong learning: Shaping inclusive and responsive university strategies*. Brussels: EUA Publications.

Sursock, A. (2015). *Trends 2015: Learning and teaching in European universities*. Brussels: EUA Publications. Retrieved from www.eua.be/Libraries/publications-homepage-list/EUA_Trends_2015_web

UKCES. (2014). *The future of work: Jobs and skills in 2030*. Retrieved from https://assets.publishing.service.gov.uk/government/uploads/system/uploads/attachment_data/file/303334/er84-the-future-of-work-evidence-report.pdf

World Economic Forum. (2015). *The Global competitiveness report 2015–2016* (K. Schwab, X. Sala-i-Martín, R. Samans & J. Blanke, Eds.). World Economic Forum (2015/2016 ed., Vol. 5). Cologny/Geneva: World Economic Forum. doi:92-95044-35-5

4

A NEW AMBIENCE OF INTERNATIONALISATION, DIGITALISATION AND SUSTAINABILITY FOR STUDENTS THROUGH PEDAGOGY

Interviews with the three universities' rectors

Carol Switzer and Marijane Luistro-Jonsson

Hanken School of Economics: An interview with Karen Spens, Rector and Maj-Britt Hedvall, Director of International Affairs

Part 1 Internationalisation

How do you define internationalisation with respect to the university's strategy? Describe your views on the importance of international student and faculty collaboration.

The cornerstones of Hanken's mission are academic research with corporate world relevance, research-based education, internationalisation, ethics, responsibility and sustainability.

In the strategy, internationalisation of all activities is embraced. During the last century, the activities in teaching and learning have developed in line with the demands of the surrounding society, in particular, the corporate world. The Finnish economy, a small open economy, is among the most globalised in the world. It is characterised by Finnish companies operating globally and international companies operating in Finland. Consequently, the students must be trained for acting in the global economy and changing society regardless of whether their future careers are in Finland or abroad. The degree programmes target national and international students and include international content and international learning experiences in order to prepare them for a career in the global academic community and corporate world.

The level of the faculty's internationalisation, the professors above all, ensures Hanken's ambition to be a research-oriented university-level business school acting internationally in research and education. Faculty members are engaged in international research networks and a significant share of the publications are internationally co-authored. The faculty consists of domestic and international core faculty members and international vising faculty. At present, a quarter of the core faculty is international. No more than a third of the core faculty can be international with reference to the national responsibility for educating business graduates fluent in Swedish.

Hanken has fully adopted the concept of 'teaching and learning in the international classroom', which can be physical or virtual. International degree and exchange students do not take separate courses designed especially for this cohort, but participate in the same courses as the domestic students. In addition to mixing international and domestic students, the international classroom consists of Hanken's own international faculty and domestic faculty with international experience, visiting faculty, guest lecturers from the international corporate world, course literature that is almost entirely international, and international cases and assignments.

Does the form of internationalisation offered by The World in the Making: Tackling World Challenges (TWC) course have merit as a valid alternative?

Hanken places special emphasis on internationalising its domestic student body. Hanken is the first, and so far only, university in Finland to introduce a mandatory semester abroad for its bachelor's students. In addition to this, the mobility scheme consists of internships abroad, language courses abroad, summer schools abroad, course study tours abroad, studies abroad in connection with the master's and doctoral studies, and more recently the virtual international classroom.

The TWC course is Hanken's first virtual international classroom. The course is assessed as a success and it is truly in line with Hanken's international ambitions. The course enables Hanken to offer an international learning experience to its MSc students without having to transport them abroad for a whole semester, and the students work virtually in intense international teams, which resemble the way global businesses work today. The TWC course also exposes students to the situation of working on a project with faculty and students they have not met before, with a limited time to finish the project. Rather than producing a quick answer, students explore how to approach an unsolvable problem, a more realistic experience for them.

How much should the experience be structured? How much faculty involvement?

Positive aspects of the TWC course are the emphasis that was made by the faculty in developing a didactic environment that meets student needs. It is structured (the learning 'abroad' is set by a team from involved universities) and the methods for approaching problems are explored across themes that are aligned with the university strategy. The faculty involvement is critical – the concept was generated from research collaborations which underpin the strength of the course.

Part 2 'Wicked problems' or World Challenges

The course tackles problems that are important to sustainability.
What is the university's strategy on approaching these issues?

Along with internationalisation, sustainability is a cornerstone in Hanken's strategy. Since 1994, Hanken has selected areas of strength in research. An area of strength is defined as an area where the research conducted is at an internationally competitive level. These areas are prioritised when deciding upon investments in research and are selected by the Board of Hanken for a period of five years at a time. Responsible Organising was chosen as an area of strength in the last evaluation of research. Responsible organising, consisting of close to 40 researchers from five subjects, focuses on how different actors, such as firms and other organisations and networks, organise for transformative action towards sustainable outcomes.

In teaching and learning, Hanken applies the Assurance of Learning (AoL) system, which establishes specific goals for integrating sustainability as a theme into student learning. This is a long-term process, since determining how sustainability can be 'measured' was not obvious initially. Now, as an example, we have incorporated criteria that consider sustainability in the evaluation of student theses. Still, this is quite difficult to measure, however, having the criteria makes the objective clear to students and encourages them to approach and resolve these kinds of issues much more than previously.

Hanken was the first university in Finland to be accepted as a member to and to sign the Principles of Responsible Management Education (PRME). The PRME initiative is the first organised relationship between the United Nations (UN) and business schools. The mission of PRME is to inspire and champion responsible management education, research and thought leadership globally. Our approach at Hanken has been that rather than have a separate course on sustainability, we address it in all aspects, in all activities in the university. Hanken offers a study module in Corporate Responsibility (CR), which is designed to provide anyone residing in Finland and interested in CR with an in-depth understanding of issues related to CR from different stakeholder perspectives. We established the CR module in the bachelor programme in such a way that students can pursue their choice of study and include the CR module as a minor. This method gives students an incentive to study CR and results in students opting to research sustainability topics in their theses. It also promotes research among faculty into areas of sustainability as they can explore it with students during their teaching.

Perhaps most importantly, it emphasises the importance of this field of study since it highlights the university's commitment to the topic. It is a bottom-up approach rather than top-down and depends on the spirit of the faculty and the students to investigate these areas. It has been successful at Hanken.

Is the course advancing the university's strategy? Is there more that the course should be doing to align with the university's strategy?

The TWC course is advancing our strategy and is aligned with it, since the strategy emphasises research-based education providing graduates with an ability to think analytically and critically in order to act in the continually changing environment. The course is also a part of the evidence that supports our goals in education on sustainability in a global context. Approaching wicked problems is appropriate at the master's level. Most of our MSc students have their bachelor's degree from Hanken. Now, with CR well established in their bachelor's studies, they have a solid background to refer to in the TWC course when aiming to solve wicked problems with no quick and easy solutions.

Part 3 Working with global companies

How important is it to involve global companies on tackling large challenges such as food waste?

Our assumption is that the TWC course gives greater access to tackle the large challenges in cooperation with global companies since, for a global company, it is more appealing to work with universities from three countries rather than one. It improves our ability to attract global corporate partners and we welcome the exposure the students gain as a result.

Part 4 Digital technology

Digital technology has developed organically at Hanken. With activities in two cities in Finland 450 kilometres apart (Helsinki and Vaasa), we adopted digital technology early as it was feasible to avoid travelling. Originally, online solutions were developed to give students in one location access to courses in the other. In addition, it was assumed that digital solutions would make the university more effective. Working virtually is nowadays a day-to-day activity for faculty, staff and students in both locations.

Describe the university's position with regard to investing in technology for the classroom? How does the university balance the emphasis on technology vs. content?

All courses at Hanken are content-led, not technology-led. Nevertheless, connecting with digital technology is second nature to everyone at Hanken since we have been doing it for such a long time. There has never been a specific mandate from the university to pursue it in the classroom, yet any member of the faculty who wishes to develop a course or other activity with digital elements is supported and resources are provided. A faculty member who develops an online course does not need to know the details of how to put it online since we have support staff to make it work. We have a Teaching Lab that focuses on digital solutions and is equipped to support innovations in this area.

The international collaboration at support staff level has been an unusual and a very positive aspect of the TWC course, e.g. the IT teams working together across universities in developing the technology, sharing knowledge about platforms, and strengthening the partnerships between the universities including staff visits between universities.

Part 5 Pedagogy

The TWC test a new form of pedagogy in the classroom – the Flipped Classroom, where students lead and explore subjects followed by theory. Another pedagogical innovation is the partnership across universities within the class – each team combining students across the universities. What do you think about this approach in higher education?

Innovation is important. We support people who have the drive and ideas for new ways to approach challenges. We are impressed by how the TWC faculty have worked together to find the optimal mix of innovation and structure to challenge today's students. We consider the TWC course as our 'Crown Jewel' in innovative pedagogy in bringing the virtual international classroom to Hanken by applying the Flipped Classroom approach.

Can the TWC style course be scaled up within the university?

We would support scaling up this format although it needs particular investments, not only in monetary terms but also in time devoted by faculty and support staff to developing the initiative. The academic partnerships and the company collaborations are long-term investments requiring dedicated faculty and support staff interested in continuously managing and developing the relationships and the content. The TWC course works due to a long-standing and trusting relationship between research partners. This is an ideal starting point for such an innovative project, and new initiatives should also be based on long-term and trusting relationships preferable in research.

We are willing to continue to develop and explore new ideas with both the TWC course partners.

We have a long history of collaboration with the Stockholm School of Economics and the TWC course benefits from this. We are very pleased with the new types of collaboration with the Università della Svizzera italiana, in particular, the ICE programme (bachelor level Intercultural Communication and Economics).

Stockholm School of Economics: An interview with Lars Stannegard, President

Part 1 Internationalisation

How do you define internationalisation with respect to the university's strategy?

Describe your views on the importance of international student and faculty collaboration.

Internationalisation is something that has become a characteristic of life in general, and business life as well. For Sweden, it is important to have some kind of international understanding and international outlook since it is a small country, with a small language and small home markets. So, being exposed to different types of international elements is something that is absolutely key for all of our students.

When we speak of internationalisation, we basically mean that we want to internationalise here on campus, in Sweden. We are not interested in setting up campuses all over the world. We already have two: one in Riga and the other in Russia. We want to internationalise on site here on campus. However, we do not want to make this place completely international because that would potentially lose the connection to its Swedish roots. So, what we want is to be internationally Swedish which essentially means that we are rooted in the Swedish context with a variety of individuals from different countries, 50% Swedes and 50% internationals. This is international enough for the Swedish students and Swedish enough to be interesting for the international students. (If it is 100% international, it will be like an airport terminal or a place that is floating around so we might as well put the school on a cruise ship that goes around the world.)

To achieve this, we need to develop cultural understanding and cultural literacy among the students. I believe one of the key competences of the future is to develop empathy, to understand other people, to put yourself in their clothes or walk in their shoes. If you do that, it is a competitive advantage against the robots, artificial intelligence and machines, because trying to understand others is of utmost importance. It helps you become a better marketer or leader or financial analyst – also, of course, a better individual. Cultural literacy comes from internationalisation, and that is a way to develop your empathy, and empathy is one of the key knowledge needs of the future. Empathy is the basis of understanding people from other cultures; you are not able to do that if you are not used to an international context. One will simply not be fit for future working life, or life in general. The formative years during bachelor and master studies are a good age to develop cultural knowledge.

This type of composition and interaction is the same for faculty and staff. Being homogeneous makes you short-sighted and lack understanding of the world. I would like to see 50–50 (between Swedes and internationals) there too.

Part 2 'Wicked problems' or World Challenges

The course tackles problems that are important to sustainability.
What is the university's strategy on approaching these issues?

Sustainability is one of our focus points – FRIS (Finance Retail Innovation Sustainability) and Misum (Mistra Center for Sustainable Markets) have been substantial initiatives of the school in recent years. A business school not focusing on sustainability is going to be or is already completely redundant. It is a moral obligation for universities to try to come to grips with the problems that we are facing, but also a preparation of how the world looks since we need to find new solutions for the problems that the world is seeing. We need to shift our mindset and understand how the world functions. If you do not have sustainability and the global challenges – if you do not understand the implications and what the global challenges actually are, then you will not be fit for the future. It is of absolutely crucial importance. Going into a business that does not care about sustainability is going to run out of business – it is as simple as that.

Is there more that the course should be doing to align with the university's strategy?

SDGs and sustainability issues should be fully integrated into perhaps not all the courses but in all the programmes. I am not sure if every single course needs to have sustainability elements, but there need to be different types of connecting courses, questions, problematisations contextualisation, to really understand what sustainability means.

If you study economics for instance, you need to know the basics – you need to know the language of economics, the language of finance and understand the basis of economic theory. If you do not understand the basis of the theory, you cannot just come and say that everything needs to be sustainable. It doesn't really work. You need to know the basics and the facts, and then you add the contextualisation, problematisations, questions, asking what does this really mean? What are the implications for the planet, the individual companies and for you? We are doing this with the Global Challenge course package in the bachelor's programme but it needs to be developed more in the master's programme. It is hard to know exactly how we can do that. It is always a question of how you can get the basics and the facts together with the mindset of understanding the context of the planet.

We are also active in PRME and training the trainers, trying to educate the faculty and trying to understand the implications. We are champions in the area of PRME and are working a lot with it. Misum is developing it. It is extremely helpful – a framework that helps us to be more systematic.

Part 3 Working with global companies

How important is it to involve global companies on tackling large challenges such as food waste?

If you have gigantic companies that are responsible for much more CO2 emissions, then when it comes to food waste, the implications are enormous. I also think that focusing on big business is the right thing to do but it is also good to focus on local and small-scale initiatives because they serve as examples and also, it becomes possible to understand at the micro-level how things can be organised. It will just be a matter of scale afterwards. I think, of course, it is super necessary to focus on the big organisations but also interesting to focus on the small ones.

We have 110 corporate partners. What they are basically all asking for – and that is why we are focusing more on sustainability issues – are questions connected to the SDGs in one way or another. Almost all of them want and need creative, analytical and emotionally mature individuals who are able to operate in an international context, who really understand these big problems. If you are able to come up with creative solutions, when it comes to food waste for instance, the students are interested in doing it and the companies have many stakeholders including shareholders where reducing food waste could have an enormous economic upside if done in the right way (e.g. if through service rather than products). So, I see a great demand from the companies and our corporate partners for individuals who can solve problems like these.

Part 4 Digital technology

Describe the university's position with regard to investing in technology for the classroom.

The digital strategy of the school is very simple. We want to have the latest and best available technology to improve our pedagogy. We don't want our digital equipment be an end goal in itself but a tool for learning. I actually believe that it is not that complicated anymore – what you need is basically a really good connection. We need to have technology that maximises learning and opens up for new ways of interacting. For me, it is a lot about how to actually increase learning and how to elevate the quality of the interaction – how do you use each other's time in the best way.

How does the university balance the emphasis on technology vs. content?

It is always about the content and to what extent does technology facilitate that. I do not want to be technologically fixed, let us say we will work on AI learning models (i.e. it can help you out to find the best learning paths) – that could possibly be a remarkable addition to the student's learning but it is not going to replace the interaction itself. So it is always about content – how do you teach and how do students learn – and the best way for technology to help you. So, it is much more content, and learning-driven than technology-driven.

Why should you push a particular technology if they already have something that is working for them? Technology serves to increase learning and make interaction friction-less: content before technology.

Part 5 Pedagogy

The TWC tests a new form of pedagogy in the classroom – the Flipped Classroom, where students lead and explore subjects followed by theory. Another pedagogical innovation is the partnership across universities within the class – each team combining students across the universities. How does SSE deal with the variations among the schools?

If you would expand this course as a cookie-cutter thing, where everything is the same, the way you teach and the way others teach – then, the *raison d'être* of the entire course is gone. You have to open up to individual differences. It is also a way of teaching and learning different styles – something that shakes things up – this is how they work and how I work, so how do I actually blend in and how do I operate in that type of environment? How do I learn, and how do I contribute to others' learning in that environment? It all goes back to cultural literacy and understanding diversity. Diversity is not just internationalisation but it is also about mental models and ways of viewing the world and ways of working. Diversity should not be confused with being too simplistic – it is about all of that too (teaching models, mental models).

What do you think about this approach in higher education?

I think it is very good. A good higher-education institution should teach the student the facts. We have this educational mission called FREE that represents an explicit educational mission in the school (that we did not have for the past 100 years). It is based on the fact that the world is changing with digitalisation, the geopolitical landscape looks different, and we need to adapt. Good education is about liberating the mind, liberating the individual and opening different types of avenues for their walks in life. Technology is very much part of our mental models. The acronym stands for:

(A) F – Fact and science-based: you have to understand theory and be able to distinguish between different knowledge and knowledge claims in a world of Cambridge Analytics, fake news and deep fakes where people are trying to change your view of the world. You need to be very analytical, very fact-based. That is the basis of everything, including sustainable development.

(B) R – Reflective and self-awareness: understanding who you are, how you react to others, and constantly reflecting. If you reflect, that is sort of a vaccination against mental models that become obsolete or knowledge that no longer holds true. To be reflective is incredibly important.

(C) E – Empathetic and cultural literacy.

(D) E – Entrepreneurial and responsible.

The FREE principles capture the mindset of what we need to have. Traditional higher education is very much focused on the F – facts and science-based.

However, if you increase your ability to reflect, if you become empathetic and culturally literate, and if you train your creativity and develop responsibility, then, that is the insurance of what the world needs.

Experiments in pedagogical models are basically about cultural literacy, they are basically about creativity. If you are used to facts – in university we teach where there is theory, then you study and take your exam, while the TWC idea is obviously much more innovative, explorative and Flipped Classroom, and we just have to cope with that and make the students understand. It is up to professors to make the students learn as much as possible and if they expect to be served everything, perfectly and neatly and nicely cut on a silver platter, that is not what higher education should do – that's what high school studies are. So, I think this is an essential part of our education.

Can the TWC style course be scaled up within the university?

Yes, probably. Since technology is developing and also our proficiency in using technology for conveying interest in context is increasing, I think it should be possible but I am not sure how – it should be up to you [professors].

Università della Svizzera italiana: An interview with Boas Erez, Rector

Part 1 Internationalisation

How do you define internationalisation with respect to the university's strategy? Describe your views on the importance of international student and faculty collaboration.

Internationalisation is key to USI's strategy. Universities are by nature and mission open to international exchange. Our university includes some 65% non-national students and an equivalent ratio of non-national faculty, making us the 'most international university in Switzerland' so it would actually be rather strange not to have international collaborations among universities. This said, and given the importance of the national grounding of higher-education institutions (HEIs), it is not always easy to promote international collaborations the way we should.

Does the form of internationalisation offered by The World in the Making: Tackling World Challenges (TWC) course have merit as a valid alternative? In what ways?

I think TWC is an interesting programme in which to foster international collaboration among the partner HEIs, and I'm glad that USI is part of it. My feeling is that it is in line with other important initiatives, such as those promoted by the European Commission's Erasmus Mundus programme.

Part 2 'Wicked problems' or World Challenges

The course tackles problems that are important to sustainability.
What is the university's strategy on approaching these issues?

Sustainability has finally been ratified in the university's strategy. Our approach to the issue is multifarious. On the one hand, we shall expose all of our students to the issue of sustainability at the teaching level, and on the other, we will build on our faculty's competences to offer dedicated courses on the topic. At the institutional level, we shall seriously try to 'walk the talk' and review our processes and organisation to be in line with sustainable development.

How is the course advancing the university's strategy?

It is part of the overall picture we are now drawing. Without initiatives like TWC, the university would not have been in a position to even envisage setting up a strategy, which should bear fruit in the near future.

Is there more that the course should be doing to align with the university's strategy?

Not for the time being. In the future, when more initiatives take shape, TWC will have to interact constructively with them.

Part 3 Working with global companies

How important is it to involve global companies on tackling large challenges such as food waste?

We are more and more open to cooperation with companies. Global companies are of interest to us because of their broad experience. It is of great importance that our students become acquainted with the ways such big players make their choices and use their potential. The best kind of cooperation is working together on questions and problems formulated for a common benefit. I hope that the companies in TWC will become long-term partners and continue to work like this with us.

Part 4 Digital technology

Describe the university's position with regard to investing in technology for the classroom?

We have a service dedicated to e-learning and teachers are encouraged to adapt their approach to the evolving environment. In my view, the university offers a variety of technological devices, which for the moment is sufficient to respond to teacher and student demands.

How does the university balance the emphasis on technology vs. content?

I would hope that the importance of technology never takes over that of content.

Part 5 Pedagogy

The TWC test a new form of pedagogy in the classroom – the Flipped Classroom, where students lead and explore subjects followed by theory. Another pedagogical innovation is the partnership across universities within the class – each team combining students across the universities.

What do you think about this approach in higher education? Other challenges for pedagogical advancement you deem important?

I think it is an interesting approach. It will be important to evaluate it after some time, to see if it has been an efficient choice.

Can the TWC style course be scaled up within the university?

At USI, there currently is and should be a varied approach to teaching, depending on the disciplines and the overall objective of the courses. So the success of TWC will certainly serve to show colleagues that its approach is viable, and can thus serve as a possible good practice.

5

TACKLING WORLD CHALLENGES THROUGH RENEWED CONSUMPTION HABITS

Alexandre Grandjean

Introduction: welcome to the 'Anthropocene'

According to current academic and militant debates, we – as humankind and planet altogether – have entered into a new geological era: the Anthropocene. In recent years, this term has acquired the status of an emblem for much of the spectrum of contemporary ecological movements ranging from 'ecological modernisation' paradigms to radical options such as those inspired by Arne Naess' 'Deep ecology' (White, Rudy, & Gareau, 2016, pp. 35–50). In any case, through the concept of 'the Anthropocene' and its variations,[1] the negative impact of human activities on the biosphere and geosphere are publicly acknowledged. International conferences, museums and notably local wine-crafters have gathered around this notion. For instance, and as an introduction to my following observations and arguments, Jean-Christophe Piccard, an inde-pendent wine-crafter experimenting with 'natural wines' from the region of Lavaux (Switzerland) has named one of his cuvees by this name. The bottle features on its label an image of a bunch of grapes graphically showing a rise of temperature as often represented in prevision maps of global warming. On his website he describes this bottle under the idea of recognising the agency of human activities on the planet:

> The principal theme of this exceptional cuvee is setting forth this fundamental principle: 'human activities INFLUENCE the evolution of our Planet and it is time for us to take RESPONSABILITY'. (Piccard & Daley, n.d.)

This notion of evolution of the planet is one strong idea of the agronomy he is working with to care for his soil and plants: biodynamic agriculture. Interestingly, in this holistic version of the 'organic' farming movement calling for more 'naturalness' (Barton, 2018; Besson, 2011), acknowledging human responsibility could also potentially be reversed in considering positive effects of human activities, not solely on human interests by also on enhancing and creating wider, resilient and balanced ecosystems.

In contemporary debates over the Anthropocene, the position of French philosopher Isabelle Stengers is noteworthy. According to her, another set of questions are more urgently in need of an answer than whether we have entered a new geological era or not. One of these is notably: 'How can we get out of the so-called "Anthropocene" as quickly as possible, before its cascading consequences ... irremediably destroy our world?' (*my translation*, Stengers, 2017, p. 7). Following her call, the critical situations humankind are confronted with can be envisioned as vectors of unexpected and yet promising quests towards new economic, political, technological and ontological paradigms. These would enable us to tackle current world challenges, such as sustainable energy transition and uses of resources, promotion of new kinds of biodiversity, new ways to meaningfully frame human production and consumption habits (Davies, Fahy & Rau, 2014), wealth distribution, as well as reduction of greenhouse gases and 'micropolluant' emissions in the atmosphere and waterways (Chèvre & Erkman, 2017).

I shall argue here through ethnographic observations and anthropological considerations that the 'drinking culture' of wine (Wilson, 2004; Demossier, 2004, 2011), and especially some of its radical margins composing the strand of 'natural wines', is providing inspiration to tackle these challenges: first, by subverting conventional 'tastescapes', economical practices and codes of social interactions; secondly, through new production and consumption habits and attributes. Indeed, as we shall see through a brief ethnographic vignette, the 'natural wine' milieu is promoting counter-cultural values of conviviality (Illich, 1973), care and intimate awareness for the environment as well as aesthetic and sensitive evaluation of quality and worth (Fourcade, 2011; Beckert, Rössel, J. & Schenk, 2017). In continuity with this vignette, I shall develop three ideal-types in which human consumption can be reflexively framed: *ingurgitating, swallowing* and *incorporating*. Each suppose different ways of engaging in (un) poetic ways of *dwelling* in the world and constituting specific (or restricted) relations with what surrounds us, locally and globally. Alongside the text, the reader shall find some pedagogical toolkits to animate workshops, meaningfully interrogating our consumption habits with regard to wine and any other product.

The birthplace of 'natural wine' in France: the 'Crieurs de vin'

In the medieval city centre of Troyes, a capital of the French region of Aube, one wine bar is worth our attention: the 'Crieurs de vin'. When entering, one is greeted by shelves and boxes filled with wine bottles in a barely organised manner. The wine *connoisseur* – the 'label drinker' as they are often mocked – would not find any of his usual cuvee from Burgundy or Bordeaux nor world-celebrated Chateaus and 'terroirs'. On the contrary, most of the exposed bottles are from commonly low-rated vineyards[2] and they all address minimalistic graphics on their labels, either self-made or featuring caricatures by famous left-wing French comic artists. In the same manner, the walls are filled with previous wine festival posters also made by these comic artists. Notable, one can read slogans such as 'Seek the natural, it will always whip back!' (*Chassez le naturel, il reviendra au goulot!*) or 'Learn the gesture that saves wine-crafters' (*Apprenez le geste qui sauve les Vignerons*). Humour, satires and calls to 'ivresses' are part of this culture of subverting codes from mainstream and aristocratic wine culture (Château, 2009). In a way, it engages in a 'carnivalesque' and Bacchic modality such as famously studied by Mikhail Bachtin (1984).

Most of the bottles have no certification labels or Appellation of Controlled Origin, the only *ad minima* mention of the 'Vin du pays' and their outlook makes it clear for those who have been exposed to these codes that these bottles are 'natural wines': namely wines made out of grapes in which no chemical treatment (herbicides, pesticides, mineral fertilisers, etc.), and almost no oenological inputs have been employed throughout the process. This wine is often framed as having the 'authentic taste' before the emergence of oenological sciences throughout the 20th century (Citerne, 2012). Interestingly, any attempt to formalise what stands as a 'natural wine' is highly controversial, as making 'natural wine' is more often envisioned as a posture than as strictly formal specifications (Pineau, 2019, pp. 21–26). A main point of 'natural wines' is to counteract standards and classification so as to let the 'unexpected' arise from a drinking experience. Their prices, though more expensive then entry level wines, range from 8 to 20 euros,[3] and up to now, is not subject to much speculation, contrary to wines such as those from Romané-Conti (Burgundy) whose selling is highly restricted and ranges between hundreds and thousands of euro.

The shopkeeper on site of Crieurs de vin founded it in the 1970s and it was one of the places in which the 'natural wine' phenomena started in France before it spread to nearly every 'global city' (Sassen, 1991) in recent years. After exchanging a few words, he grabs a bottle and offers me a taste of wine while continuing the story of what makes this place 'special'. He explains how he doesn't care for any certification and prefers to personally know the producers and follow them at every stage of production. Contrary to 'standard' wine retailers, he purchases bottles from them disregarding good and bad years of harvest, as he believes in practising solidarity with them during harsh years and celebrating with them for exceptional seasons. I was accompanied by a friend organising oenotouristic courses in Switzerland. After he explained his activity, and I mentioned that I am a wine anthropologist, the shopkeeper decided to show us his cellar and his wine collection. Displacing a few boxes of bottles, he uncovered a trapdoor that led us directly under the historical foundations of the building.

The scenery of this two-floor ancient cellar immediately activated my imagination. As we entered and discovered this 'hidden place', I felt as though every stored bottle was similar to seeds, waiting underground for the proper germina-tion time. Through an exercise of imagination, these bottles were not only raw grape juice having been fermented but were the depositaries and archives of times and years that had vanished. They condensed the interaction of soils – their micro biotic activities –,vine root systems and their symbiosis with mush-rooms (mycorrhizas) so as to metabolise minerals into nutriments, as well as the gestures and worldview of wine-crafters and their employees throughout the years. All these were stored in a bottle, still evolving as no antioxidants were used to stop the fermentation process. These archives were 'living' memories for those who had been previously familiarised to consider them under such an aesthetic. Some bottles were stored for more than 20 years, and the shopkeeper was eager to contradict those who maintain that 'natural wines' do not age well. Learn about the influence of basic inputs on taste in Pedagogical toolkit 1, Box 5.1.

BOX 5.1 PEDAGOGICAL TOOLKIT 1: TASTING THE INFLUENCE OF BASIC OENOLOGICAL INPUTS (SUGAR)

Acquire three litres of the cheapest red wine you can buy in the local super-market. Insert the wine in three masked bottles. In each bottle add in two, four and six spoons of plain sugar and mix it well. Have your audience taste all three bottles and discuss different tasting properties as if they were three different wines. What specific aromas arise from each bottle? Which one would they rate as the most interesting one?

Explain the trick to your audience. Wine in general is rarely made of only raw and fermented grapes. Oenological sciences have brought numerous toolkits to winemakers to 'design' wine according to specific standards. These inputs can range from wood chips to gelatin, fish glue, dried yeast, sulphur and carbon dioxide (SO_2, CO_2) and have different purposes (taste, texture, etc.). Numerous websites provide lists of authorised inputs and vinification technics according to national certification standards. According to your audience: What are the limits and practical negotiations one is willing to employ if they were to produce wine? Are some wines said to be 'natural' because of the production process they entail or in counter-relation to a dominant wine market assessed as 'technical' and eventually 'artificial'?

Where the wine-crafting radicals meet: wine-swapping and conviviality at the MABD annual congress

I was in Troyes to study the social dynamics within the 2018 annual congress of the MABD (the French Movement for Biodynamic Agriculture) dedicated to wine-crafters. Though not all 'natural wines' are being cropped according to the guidelines of biodynamic agriculture, this agronomy represents one emblematic strand of the movement towards a 'living nature' (Choné, 2013, p. 24) and a quest for 'naturalness' (Pineau, 2019). This esoterically driven agronomy, inspired by the charismatic founder of the anthroposophical movement, Rudolf Steiner, is adapted and translated by wine-crafters to tackle contemporary challenges. Similar to 'organic' farming, these wine-crafters refuse to administer any agrochemical treatment products, and rather employ 'natural' solutions to 'care' for their plants such as herbal tisanes, enhanced composts or specific biodiversity management practices. They are distinct from standard 'organic' farming methods through the use of 'biodynamic preparations' made of cow manure and crushed quartz (silica) fermented in cow horns buried for six months. These preparations, fostering an alchemical imagery, are said to give 'indications' or 'impulses' to plants and soils to balance a cosmic order made of earthly and cosmic 'forces'. These preparations are conceived as 'healing' practices following the mainlines of alternative holistic therapies (Koch, 2015; Levin, 2008). Relying on (supra)sensitive forms of knowledge, this esoterically driven agronomy is challenging contemporary scientific models to cropping plants (Compagnone, Prévost, Simonneaux, Lévite, Meyer & Barbot, 2016).

During the congress, hundreds of producers came from every region of France, plus a few from Switzerland, Italy and Catalonia. In a sense, they represented the radical and 'underground' scene of western European wine-crafting. This scene features those who are willing to address and confront the environmental and economic 'risks' (Beck, 1992) potentially losing part or their entire yearly production by not allowing themselves to secure their harvest with technical 'artifices' such as chemical pesticides or fungicides. This feature is notably why their methods can be described as radical (Pineau, 2019). The congress offered lectures from 'alternative' agronomical experts, commentators on Rudolf Steiner's writings, and wine-crafters sharing know-how and experiments they had conducted on promoting local biodiversity, producing 'green fertilisers', working with animals in the vineyards and learning to manage 'spontaneous fermentation' with yeast naturally present on grapes. During lunch and dinner, every wine-crafter brought their bottles which were shared from table to table, hence constituting a panorama of European 'alternative' production. Following a 'sensory ethnographical' approach (Pink, 2015), it is interesting to note that the typical tasting dispositions were not followed. Wines were hardly detailed with formal organoleptic attributes, but rather described through subjective notions such as feelings, impressions, memories or narratives of events that occurred during that year of cropping grape.

White and sparkling wines followed red wines with the Riesling from Alsace and its typical taste of petrol blended with other tannic wines from Languedoc, some featured so-called 'defects' such as phenol or ascetic tastes, others were unfiltered – looking cloudy – or seemed to be merely grape juice with flavours enhanced by alcohol. Other wines were more 'standardised' depending on the producer's attitude towards experimentation and discourse over 'terroir'. Following the Bacchic attitude of subversion enacted during these bottle swapping moments, it was important for wine-crafters not to enact the common hierarchies of the mainstream wine market, but on the contrary finding commonality between each other (building a *we*-group). Indeed, wine-crafters I was sitting with had an attitude of promoting and agreeing together on what they would call the 'expression' of their vines, of their soils – hence detailing the geological composition of the latter – as well as how their plants were reacting year after year to climate variation. Parallels were also made between their 'inner selves' – how they would relate intimately to their lands and local ecosystems – and how their wines would evolve and change over years. For some of them wine was a 'soul product', connecting to the realm of the spirit as one wine-crafter especially described moments of almost collective ecstasies in which 'aegregores'[4] manifested themselves while they were tasting bottles blindfolded.

In the evenings after lectures, we would all gather in the 'Crieurs de vin'. Once again, bottles were to be shared from table to table. Stating that rules of tasting conventions were to be disrupted, the music played was 'free jazz', a music that indeed fits the overall ideology and representation on which one always has something to discover about himself or herself while encountering new wines and a new spectrum of tastes. Similar to experimental music, the 'natural wine' milieu also stresses the importance of learning to 'unlearn' the conventions. This carnivalesque feature made me think over Arjun Appadurai's theory of global flux (1996). To him, globalisation has to be understood through the mobility of different scapes. Overall, he describes and considers five types of scapes: ethnoscapes (mobility of human populations), mediascapes (mobility of images and social imagery), financescapes (mobility of capital), technoscapes (mobility of technology). Considering the strong 'sensory' and embodied aspects of every human being and 'material culture' scholars' renewed interest (Berger, 2009; Morgan, 2010), it is also important to consider what is at stake with 'tastescapes'. Namely, mobility that is conferred to how one product should taste or (not) resemble while being incorporated in a sensory human body in specific contexts.

Take for instance Thomas Wilson's (2004) example on how the 'Irish pub culture' has spread globally, say from Galway and Dublin to Barcelona and Jogjakarta. Anyone who has already visited such 'pubs' easily grasps and is reminded of the specific 'tastescape' associated with Guinness[5] and its overall 'ambiance' constituted by shady wooden bars and Celtic music. Now, in the case of 'natural wine' culture, these 'tastescapes' bear an essential attribute: they subvert – sometimes even to the extent of pastiche – standardised tastes such as the worldwide blend of Merlot and Cabernet-Sauvignon coming from the ideal type of Bordeaux. If one has turned global (Ulin, 2004), and still bears many variations across country of provenance, on the contrary 'natural wines' are about details, surprises and most especially ultra-localness. In a sense, 'natural' is only what does not resemble the rest of the production, stressing emphasis on nuances and variation year after year. Learn about tastescapes in Pedagogical toolkit 2, Box 5.2.

BOX 5.2 PEDAGOGICAL TOOLKIT 2: ENCOUNTERING THE TASTESCAPES OF A WINE

Ask a professional wine-retailer to introduce you to a local wine-crafter he is currently promoting. Ask whether s/he has documentation on the wine-crafter (pictures of vineyards and cellar, specific working technics, videos, etc.) or ask if you can record the wine-retailer's own presentation on the wine-crafter. If you personally know a wine-crafter, pay a visit to him/her.

In front of your audience, have the bottle of this wine-crafter to taste without any comments. Then present the documentation and explain the peculiarity beyond the organoleptic propriety of the wine itself. How does the wine-crafter perceive his/her profession, relationship with the plants, 'terroir' and patrimony? What makes his/her wine special according to the wine-crafter? How does it constitute a specific landscape in the region that differs from other wine-making regions?

Have the wine tasted again. Then engage your audience to reflect on the importance of images and narratives on their tasting dispositions: how different is this exercise in regard to mass-consumption advertising? Does tasting local wines influence how your audience now perceives the region they live in? Do they have memories of specific places in regard to wines they have already tasted (say, during holidays or specific moments of their lives)? How many dimensions does drinking wine have? Try to collectively enumerate them and detail them.

Tackling World challenges through renewed consumption attitudes: ingurgitating, swallowing and *incorporating*

In recent years, one important strand of anthropological discourses have engaged considerations over ontologies: namely, *how do human beings inhabit, perceive and apprehend their environment and other non-human entities such as microbiotas, plants, ecosystems, the planet and even the cosmos?* [6] I assume that this ethnographic vignette is as any ethnographic vignette a singular and contextual illustration of this wide array of ontological postures. One could easily criticise it for being biased, non-representative of a wider wine market, or merely enhanced 'journalism' or 'poetry'. I rather conceive it as applied philosophy, hence giving us an opportunity to reflect on how human beings – bodies and mind entangled – engage in meaningful production and consumption practices.

My experience as a wine anthropologist comprehensively studying the dynamics at stake within radical figures of wine-crafting have initiated me in new conception: of tastes, ways of representing the interaction between humans and non-humans, and envisioning alternatives in valuing an emblematic product such as wine. Three years (2017–2019) of visiting, conducting interviews and participant observations with radical figures of wine-crafting[7] have enabled a way to conceive three ideal-types in which we – as human-kind – engage in relating to consumption. I shall theoretically present these here more as a work in progress than as a formalised model. So far, I assume three attitudes can be denoted in which we engage with products, and by extension how these enable us to engage with our surrounding environment: *ingurgitating, swallowing and incorporating*. It is to be noted, that these are ideal-types following a Weberian framework, hence not necessarily referring to 'lived' experiences. They nevertheless are conceptual artefacts, helping us think through complexity.

Ingurgitating: the frenetic consumption of 'hyperreal objects'

In a context of mass consumption and acceleration (Rosa, 2013), human beings of western and so-called Westernised countries have access to almost unlimited commodities. The offer in terms of knowledge, cultural products, worldviews, alimentation and other goods seemingly exceed demand in a global context (Giddens, 1986; Appadurai, 1996). Though, it depends on access to economic, technological, social and symbolic capitals, human beings can easily engage in frenetic consumption. We can drink cheap wines to get drunk for drunkenness sake (for instance, the practice of *binge drinking* in youth culture). We can have access to academic literature and never be able to compile all the literature pro-duced by a more and more cosmopolitan and publish-or-perish context. Cultural industries provide more series, movies, books and music albums one could ever have time to dedicate to. We have access to food supplies and cooking styles in an unprecedented manner and are able to experiment new creative blending and fusion, though most restaurants propose standardised alimentation. This setting provides the condition opposed to scarcity: abundance or even saturation.[8] It is nonetheless not surprising that 'voluntary sobriety' and 'slow food' movements have emerged since the 1990s to set limits and conditions to this seemingly unlimited access to industrial consumption goods (Petrini & Padovani, 2006; Rabhi, 2010).

In this context, it is easy to engage in a bulimic attitude towards consumption. These industrial goods are being 'ingurgitated' by humans turned into 'con-sumers'. They transit through stomachs, minds, imageries, yet not necessarily 'feeding' them, or providing any kind of agency to human beings. This attitude is one of being passive, not being held responsible for any side-effects of the global economic system of production and consumption of goods. Interestingly, it seems to reverse the modernist posture between object and subject. In a way, the attitude of 'ingurgitating' considers humans merely as depository objects of wider processes: transforming alimentation into excreta (not even compost), intellectual theories in name dropping, narratives and aesthetics into forgotten memories and images, etc. As philosophically posed by Jean Baudrillard (2014 [1970]), it constitutes objects of 'mass consumption' which are overcharged with meanings, symbols and presences which paradoxically neutralise their values, uses and relation towards 'reality'. The semiotic relationship and mediation between the object and its social significance is no longer guaranteed. Food then becomes solely food to 'ingurgitate' in a tautological manner: *ingurgitate, you ingurgitate, we ingurgitate because we ingurgitate …*

Swallowing: the human utilitarian approach

The first posture (ingurgitating) corresponds to what could be described as a post-modern attitude, in a sense fearing a loss of reference points and ways to relate to a material reality. A second posture can be related to a more modern paradigm to consumption: 'swallowing'. In this approach, consumption goods are perceived as part of a mechanical and utilitarian cosmology: the human is a machine that has to be 'fed' for specific purposes. The representatives of this consumption attitude are for instance dietary scientists and promoters of so-called 'super foods'.[9] The latter are aliments which are considered, disregarding their tasting proprieties, for what they provide in terms of nutrients and calories. They represent a way of thinking that translates value into units of measure. To perform fully, standards of alimentation have been set for human bodies and cognitive capacities. To be fully ethical, diets such as vegetarianism or veganism have to be pursued. To be fully socially assimilated, 'classic' cultural production has to be mastered and wine consumption is highly codified within courtesy cultures (Elias, 1969).

This perspective focuses on the sole agency of individuals: direct effects are thought without considering any wider production context. The issues brought by growing for instance soju or quinoa on a global scale and its impact on the environment and local 'insecure' populations are not necessarily understood. Anthropologist Anna Tsing (2015) provides insightful considerations on the potential overlaps of life cycles encountered in one product with her work on the social trajectories of the matsutake mushroom, at the crossroad between gatherers in Oregon and glocal retail markets. In her ethnographic work, she notably highlights how former American war veterans encounter Vietnamese and Cambodian migrants in making a living out of dysfunctional welfare states, traumatic experiences and specific skills acquired during warfare. Anna Tsing raises issues on the 'hidden transcript' (Scott, 1992) of everyday consumption goods. The latter having a long history of multi-scaled modes of production involving numerous beings (humans and non-humans). These interactions are being occulted by a representation in which individuals are 'swallowing' for their own use, without considering the numerous interconnections and frictions that arise out of these global retail markets (Tsing, 2005).

Incorporating: consuming with body, mind, imagination and others

One last consumption attitude can be detailed here with more emphasis, as it stands for 'emerging structures, norms and new beginnings' (Beck, 2016, p. 39) in line with contemporary 'holistic' consideration on the interconnection between humans and non-humans. Philosophically, it notably stands for a quest over defining a *new materialism* and *object-oriented ontologies* in reaction to postmodern ideologies, theories and neo-liberal economic practices (Bauhardt, 2013; Haraway, 2008; Morton, 2013). Practically, it addresses the issue of what – as reflexive individuals – we are willing to let enter into our bodies and minds, as well as what productive system we are endorsing. For instance, do we prefer to *incorporate* locally produced wine that defies our tasting dispositions, as well as promote environmental resilience, or do we want a wine that fits oenological standards, but not necessarily ethical ones?

By 'incorporating', I mean consumption practices that raise *awareness* on what 'feeds' us, namely being conscious of the multiple chains of interaction that enable one product to be singular, as well as provide agency both to producer and consumer. Standing against bulimic and utilitarian postures, the attitude of 'incorporation' follows the famous stance of Edmond Rostand's hero, Cyrano of Bergerac, when he states that 'it is even more beautiful when it is useless' (*my translation*, Rostand 1999, p. 417). Paradoxically, it is by not being necessarily useful and optimised that an act of consumption can provide beneficial and virtuous 'side-effects' and constitute 'accidents' that enlarge our registers of action enough to tackle contemporary world challenges.

The case of 'natural wine' previously presented can be reconsidered against the backdrop of this illustration of 'incorporating'. The radical figures of the wine-crafting milieu I described are engaged in alternative agronomies which are not necessarily following scientific crop monitoring strategies. On the contrary, they subvert standardisation through radical consideration of localness: the specific composition of their soil (the 'terroir'), local grape varieties, interaction with surrounding biodiversity, oenological procedures seeking 'authenticity' and 'naturalness', or dispensing more subjective and emotional approaches to wine sharing are part of a whole. Through the emblematic product of wine, these radical figures are telling another narrative of the world: one in which Darwin's struggle for life is counteracted by another evolutionary process: cooperation (Kroptokine, 2010 [1902]). One in which the 'perils' of modernity (for instance, the uncovered toxicity of agrochemical treatment products) can be challenged with critical models of modernity (Knauft, 2002) featuring a new understanding of 'nature', local and global, social organisation and market practices.

By drinking wine, they assume that it is a form of transubstantiated product (similar to the ritual Christian consumption during the Eucharist): in that manner, drinking wine is not only drinking fermented grape juice and worked out with oenological inputs. It is as Marion Demossier (2004, p. 2) states, 'consuming place, time and symbols'. Though this is acknowledged in most contemporary 'wine cultures', 'natural wine' promoters add to it another element: consuming the relationship and mutual work of humans and non-humans. By refusing agro chemistry, most wine-crafters engage in an ontological posture that seeks to uncover the specificities of this relationship through their wine; hence, not 'creating' tastescapes on their own but merely letting this relationship 'express' itself and constitute a complexity that can be tasted and narrated through their wines. This attitude demands the consumer to be both an active body and a mind in an environment familiarised with these new ways of tasting and relating to wines, thus engaging with imaginary processes of entities and processes that are 'invisible' to human perception senses: bacteria, yeast, symbiosis between mycorrhiza and vine root systems and so on. Film director and wine activist Jonathan Nossiter (2015) introduced me to this realm of radical interconnection and alterity in his book *Insurrection culturelle*. He notably explains, with a *mediation* through metaphors that:

> In the case of vines, the presence of synthesised molecules inhibited in a radical manner the capacity of the roots to penetrate deeply and to draw the mineral salts, which, afterward, shall be transformed into flavours. Every taste, every flavour of the wine then comes from the transformation of these mineral salts. This transformation depends therefore of the vitality of the biological life of the soil that hosts the roots. It is not the mineral by itself that can be swallowed by the root, it is the absorption of these mineral salts that needs to be turned into a mush. And it is the fauna of the underground, a brigade of little cooking chefs, which degrade the minerals into an assimilable substance by the roots. All the complexity of the taste of a wine comes from the surface of contact of vine roots with different geological layers. The more a root dialogues with the soil and the underground, the more it is enriched. (*my translation*, Nossiter & Beuvelet, 2015, p. 109)

This poetic statement about what constitutes the taste – and by extension the value – of one wine also underpins that human beings are not the only ones either *swallowing* or *incorporating* nutriments. It promotes a worldview of relational ecosystems in which plants incorporate minerals and turn them into nutrients and organoleptic proprieties (flavours) we in turn can either *ingurgitate, swallow* or *incorporate*. In a way, the attitude of *incorporating* is precisely acknowledging this complexity, but also the incertitude and serendipity that can arise out of it. *Incorporating* is acknowledging the intersubjectivity and interrelations that provide elements to *transform* our daily world and existence, and hopefully for common goods. Explore the nuances of natural wines in Pedagogical toolkit 3, Box 5.3.

BOX 5.3 PEDAGOGICAL TOOLKIT 3: THE LIKES AND DISLIKES OF 'NATURAL' WINES

Pick up bottles of so-called 'natural wines'. Start explaining what this kind of wine is about, notably through possible connections and comparison with different kind of arts and craftsmanship you and your audience are familiar with. For instance, *noise* music – starting with Luigi Russolo's (2003 [1913]) manifesto 'The art of noises' – bears similar features such as engaging the public with a not necessarily pleasant and convenient approach to music. The very notion of craftsmanship allows different products to be shaped each time in a personal and non-standardised manner (Cf. Sennett, 2008).

Have your audience taste the wine. What strikes them? Is the taste of the wine changing minute by minute? What kind of feelings does that release for them? When was the last time they purposely decided to consume something they would not necessarily be sure would fit their usual taste? Does the taste of the wine remind them of memories, or does it bring about metaphors to speak about? Ask each member of the audience to describe to his/her neighbour what feelings they have about this wine, and eventually reflexively why they feel this way.

Knowing that the 'natural wine' trend is not bound anymore to subversive counterculture, do they understand why 'natural wine bars' can be found in any major city in Europe and America? If so, what are the numerous ideologies that can be raised on promoting the consumption of 'natural wines'? Is it about social distinction, experiences of incertitude, training to be 'flexible' with one's tasting disposition, discovering and fostering so-called 'traditional' tastescapes before the rise of agrochemistry, or are ecological and sanitarian arguments preeminent?

Conclusion: how drinking (natural) wine could get us out of the Anthropocene

Following Isabelle Stengers, 'how can we get out of the Anthropocene?' is a meaningful and thought-provoking question to ask. In regard to my considerations starting with the 'nature wine' milieu and on the three ideal-types of consumption, I have shown that through an overview of the 'natural wine' environment and a reflexive approach towards consumption habits, there is a possibility to bring forth ideas of poetic *dwelling* on the planet with new embodied and imaginary models to emerge. As pointed out by Kevin Hetherington (1997), in the 18th century, multiple 'social laboratories' (heterotopias) such as the first industrial factories, the Palais Royal or the Freemason lodges have been places where the components of modern *ethos* and values have been experimented upon before becoming mainstream. In the 'Anthropocene' era, I believe the vineyards and the 'natural wine' milieu to precisely represent a new generation of 'laboratories' that experiment and actualise new modes of social order. The pedagogical toolkits provided in this text are a possible methodology one could follow to simulate these kinds of 'social laboratories'.

The counter-cultural scenery I have described of conviviality, disruption of common hierarchies, or of linking production with 'inner selves' or holistic insights constitute practices in which human activities are not only negative influences on the biosphere and the geosphere. On the contrary, through a posture of *incorporating*, we could consider how non-humans are also part of the equation to be sought. In a way, through a new understanding that brings forth models of biodiversity management, or imaginary processes, it could lead to new practices that valorise the mutual and potentially beneficial impact the anthroposphere, the biosphere and the geosphere can have on each other. It hence leads the concept of the Anthropocene as a positive process that could enhance ecosystems and provide human agency simultaneously on producers and consumers alike. This could lead to a 'metamorphosis of the world' as expressed by sociologist Ulrich Beck (2016) in his posthumous book. We could indeed follow his idea that 'because climate change is a threat to humankind, we can and should turn the question upside down and ask: What is climate change *good* for (if we survive)'? (Beck, 2016, p. 35). By that means – and stressing the survival condition – he postulates that social organisations could reverse the 'perils' of modernity into 'benefits' once again.

Ulrich Beck (2016, p. 39), as an optimistic sociologist invites us on that path, especially when he considers that:

> Climate change is creating existential moments of decision. This happens unintended, unseen, unwanted and is neither goal-oriented nor ideologically driven. The literature on climate change has become a supermarket for apocalyptic scenarios. Instead, the focus should be on what is now emerging – future structures, norms and new beginnings.

The burden is then on planet dwellers to seek, outline and experiment on these 'futures structures, norms and new beginnings' that could lead us to tackle current world challenges. In a way, changing the negative aspects of the Anthropocene into positive ones, and eventually rethinking the new utopias we want and hoping to see emerge – collectively. Drinking too many glasses of 'natural wine', and thus *incorporating* it, could be a good method to start thinking about the revolutions to come.

Notes

1 In the Anthropocene debate, one strand prefers to label it 'Capitalocene' (Moore, 2016) to stress upon Western technological and economical particularities that developed during the industrial revolution. Indeed, attributing global transformations over biosphere and geosphere to the abstract figure of humankind is misleading according to this theory. Colonisation and the acceleration of globalisation processes (Giddens, 1986; Rosa, 2013) represent unprecedented abilities by capital's stakeholders to impact the global environments that most populations of the world still do not have. Approaches such as those deriving from 'Queer ecology' (Bauman, 2015; Gandy, 2012; Morton, 2010) have also provided mitigated insights on the Anthropocene debates based on Neo-Darwinist theories. 'Climate weirding' (Bauman, 2015) has to be separated from more co-evolutive and non-normative models of interaction between humans and non-humans. The 'Anthropocene' would be an anthropocentric concept still fostering a strong distinction between 'nature' and 'culture'. For Matthew Gandy (2102) for instance, 'heterotopic alliances' develop through specific forms of biodiversity, concepts of place-making and human activities, either marginal or dominant ones.
2 Marie-France Garcia-Parpet (2014, p. 73) suggests that alternative agronomies such as biodynamic farming and claims over 'natural wines' are indeed a strategy of recognition in a market that is symbolically dominated by prestigious regions and clear hierarchies within domains. It also fosters a divide between 'technological' wines and wines of 'terroir', hence challenging common taxonomical and classificatory systems used in the wine 'drinking culture' (Douglas, 1986, chapter 10).

3 It is assumed that in every wine-crafting domain, the production cost of one bottle does not exceed 10 euros (Beverland, 2005).
4 This term derives from the esoteric milieu, standing as a materialisation of shared intentionality.
5 Which like specific and world-known beer brands, tend to guarantee to have the same taste anywhere in the world.
6 See for instance the work of Tim Ingold (2000) on *dwelling* practices and the acquisition of *sensorial skills*.
7 In Switzerland, I sampled 32 wine-crafters that are engaged in biodynamic agriculture or other alternatives inspired by 'contemporary spirituality' movements (Fedele & Knibbe, 2012) such as neo-shamanism, neo-orientalism, geobiology or New Age intentional thoughts. They were selected also due to their geographic location and different cantons standing for different agricultural politics, and religious or spiritual landscapes. Eight other wine-crafters were visited as they engaged in the 'conventional' or 'standard' organic approach to grape-cropping. Semi-structured interviews were conducted with each wine-crafter, and further participant observation and textual collection has been incorporated on specific participants of my research.
8 In a sense, we can wonder whether we have achieved or not the famous state of cornucopian abundance as described in utopian literature ranging from Francis Bacon's *New Atlantis* to Karl Marx and Frederic Engel's *Communist Manifesto*. If the answer is yes, is there any teleology – intentional objectives – to this state of abundance or is it the result of unplanned side-effect?
9 I'll let the reader scout the Internet on this topic and discover the numerous controversies that define what constitutes a 'super food'.

References

Appadurai, A. (1996). *Modernity at large: Cultural dimensions of globalization*. Minneapolis: University of Minnesota Press.
Bachtin, M. (1984). *Rabelais and his world*. Bloomington: Indiana University Press.
Barton, G.A. (2018). *The global history of organic farming*. Oxford: Oxford University Press.
Baudrillard, J. (2014 [1970]). *La Société de consommation: ses mythes, ses structures*. Paris: Denoël.
Bauhardt, C. (2013). Rethinking gender and nature from a material(ist) perspective: Feminist economics, queer ecologies and resource politics. *European Journal of Women's Studies*, 20 (4), 361–375. doi:10.1177/1350506812471027
Bauman, W.A. (2015). Climate weirding and queering nature: Getting beyond the Anthropocene. *Religions*, 6(2), 742–754. doi:10.3390/rel6020742
Beck, U. (1992). *The risk society: Towards a new modernity*. London: Sage
Beck, U. (2016). *The metamorphosis of the world*. Cambridge: Polity Press.
Beckert, J., Rössel, J. & Schenk, P. (2017). Wine as a cultural product: Symbolic capital and price formation in the wine field. *Sociological Perspectives*, 60(1), 206–222. doi:10.1177/0731121416629994
Berger, A.A. (2009). *What objects mean: An introduction to material culture*. Walnut Creek, CA: Left Coast Press.
Besson, Y. (2011). *Les Fondateurs de l'argiculture biologique: Albert Howard, Rudolf Steiner, Maria & Hans Müller, Hans Peter Rusch, Masanobu Fukuoka*. Paris: Sang de la Terre.
Beverland, M. (2005). Crafting brand authenticity: The case of luxury wines. *Journal of Management Studies*, 42(5), 1003–1029. doi:10.1111/j.1467-6486.2005.00530.x
Château, L. (2009). La Dive bouteille en France: une analyse sémiologique et linguistique de son discours et de ses images. *COnTextes*, 6. doi:10.4000/contextes.4485
Chèvre, N. & Erkman, S. (2017). *Alerte aux micropolluants: Un péril invisible*. Lausanne: Presses polytechniques et universitaires romandes.

Choné, A. (2013). Les Fondements de l'écologie spirituelle chez Rudolf Steiner. *Politica Hermetica*, 27, 15–35.

Citerne, P. (2012). *Les Métiers du vin: Histoire & patrimoine*. Portet-sur-Garonne: Loubatières.

Compagnone, C., Prévost, P., Simonneaux, L., Lévite, D., Meyer, M. & Barbot, C. (2016). L'Agronomie: une science *normale* interrogee par la biodynamie? *Revue de l'association française d'agronomie*, 6(2), 107–113.

Davies, A.R., Fahy, F. & Rau, H. (Eds.). (2014). *Challenging consumption: Pathways to a more sustainable future*. London: Routledge.

Demossier, M. (2004). The quest for identities: Consumption of wine in France. *Anthropology of Food* (S1). doi:10.4000/aof.1571

Demossier, M. (2011). Beyond terroir: Territorial construction, hegemonic discourses, and French wine culture. *Journal of the Royal Anthropological Institute*, 17(4), 685–705. doi:10.1111/j.1467-9655.2011.01714.x

Douglas, M. (1986). *How institutions think*. London and New York: Routledge.

Elias, N. (1969). *The court society*. Oxford: Blackwell.

Fedele, A. & Knibbe, K. (Eds.) (2012). *Gender and power in contemporary spirituality: Ethnographic approaches*. London and New York: Routledge.

Fourcade, M. (2011). Cents and sensibility: Economic valuation and the nature of 'nature'. *American Journal of Sociology*, 116(6), 1721–1777. doi:10.1086/659640

Gandy, M. (2012). Queer ecology: Nature, sexuality, and heterotopic alliances. *Environment and Planning D: Society and Space*, 30(4), 727–747. doi:10.1068/d10511

Garcia-Parpet, M.F. (2014). Vers une nouvelle redéfinition des vins d'appellation d'origine? In A. Cardona (Ed.), *Dynamiques des agricultures biologiques* (pp. 57–74). Paris: Éditions Quae.

Giddens, A. (1986). *The constitution of society*. Cambridge: Polity Press.

Haraway, D.J. (2008). *When species meet*. Minneapolis: University of Minnesota Press.

Hetherington, K. (1997). *The badlands of modernity: Heteropia and social ordering*. London: Routledge.

Illich, I. (1973). *Tools for conviviality*. New York: Marion Boyars.

Ingold, T. (2000). *Perception of the environment*. London and New York: Routledge.

Knauft, B.M. (Ed.). (2002). *Critically modern: Alternatives, alterities, anthropologies*. Bloomington: Indiana University Press.

Koch, A. (2015). Alternative healing as magical self-care in alternative modernity. *Numen*, 62 (4), 431–459. doi:10.1163/15685276-12341380

Kroptokine, P. (2010 [1902]). *L'Entraide: Un facteur de l'évolution*. Paris: Éditions du Sextant.

Levin, J. (2008). Esoteric healing traditions: A conceptual overview. *Explore*, 4(2), 101–112. doi:10.1016/j.explore.2007.12.003

Moore, J.W. (Ed.). (2016). *Anthropocene or capitalocene? Nature, history, and the crisis of capitalism*. Oakland, CA: PM Press.

Morgan, D. (Ed.). (2010). *Religion and material culture: The matter of belief*. Abingdon: Routledge.

Morton, T. (2010). Guest column: Queer ecology. *Pmla*, 125(2), 273–282. doi:10.1632/pmla.2010.125.2.273

Morton, T. (2013). *Hyperobjects: Philosophy and ecology after the end of the world*. Minneapolis: University of Minnesota Press.

Nossiter, J. & Beuvelet, O. (2015). *Insurrection culturelle*. Paris: Stock.

Petrini, C. & Padovani, G. (2006). *Slow food revolution: A new culture for eating and living*. New York: Rizzoli.

Piccard, D. & Daley, L. (n.d.). l'Anthropocène, Villette grand cru, Lavaux A.O.C 2017. Retrieved from www.domainepiccard.ch/article-31-l-anthropocene-villette-grand-cru-la vaux-a-o-c-2017.php

Pineau, C. (2019). *La Corne de vache et le microscope: le vin 'nature', entre sciences, croyances et radicalités*. Paris: La Découverte.

Pink, S. (2015). *Doing sensory ethnography* (2nd ed.). Los Angeles: Sage.

Rabhi, P. (2010). *Vers la sobriété heureuse*. Arles: Actes Sud.

Rosa, H. (2013). *Social acceleration: A new theory of modernity*. New York: Columbia University Press.

Rostand, E. (1999 [1897]). *Cyrano de Bergerac*. Paris: Gallimard.

Russolo, L. (2003 [1913]). *L'Art des bruits*. Paris: Allia.

Sassen, S. (1991). *The global city: New York, London, Tokyo*. Princeton: Princeton University Press.

Scott, J. (1992). *Domination and the arts of resistance: Hidden transcripts*. New Haven, CT: Yale University Press.

Sennett, R. (2008). *The craftsman*. New Haven, CT: Yale University Press.

Stengers, I. (2017). Préface. In A.L. Tsing, *Le Champignon de la fin du monde: Sur la possibilité de vivre dans les ruines du capitalisme* (pp. 7–19). Paris: La Découverte.

Tsing, A.L. (2005). *Friction: An ethnography of global connection*. Princeton: Princeton University Press.

Tsing, A.L. (2015). *The mushroom at the end of the world: On the possibility of life in capitalist ruins*. Princeton: Princeton University Press.

Ulin, R.C. (2004). Globalization and alternative localities. *Anthropologica*, 46(3), 153–164. doi:10.2307/25606191

White, D.F., Rudy, A.P. & Gareau, B.J. (2016). *Environments, natures and social theory*. London: Palgrave.

Wilson, T. (2004). Globalization, differentiation and drinking cultures: An anthropological perspective. *Anthropology of Food*, 3. doi:10.4000/aof.261

PART II

Ingredients: Blender, diversity, learning, international, encounters, broken

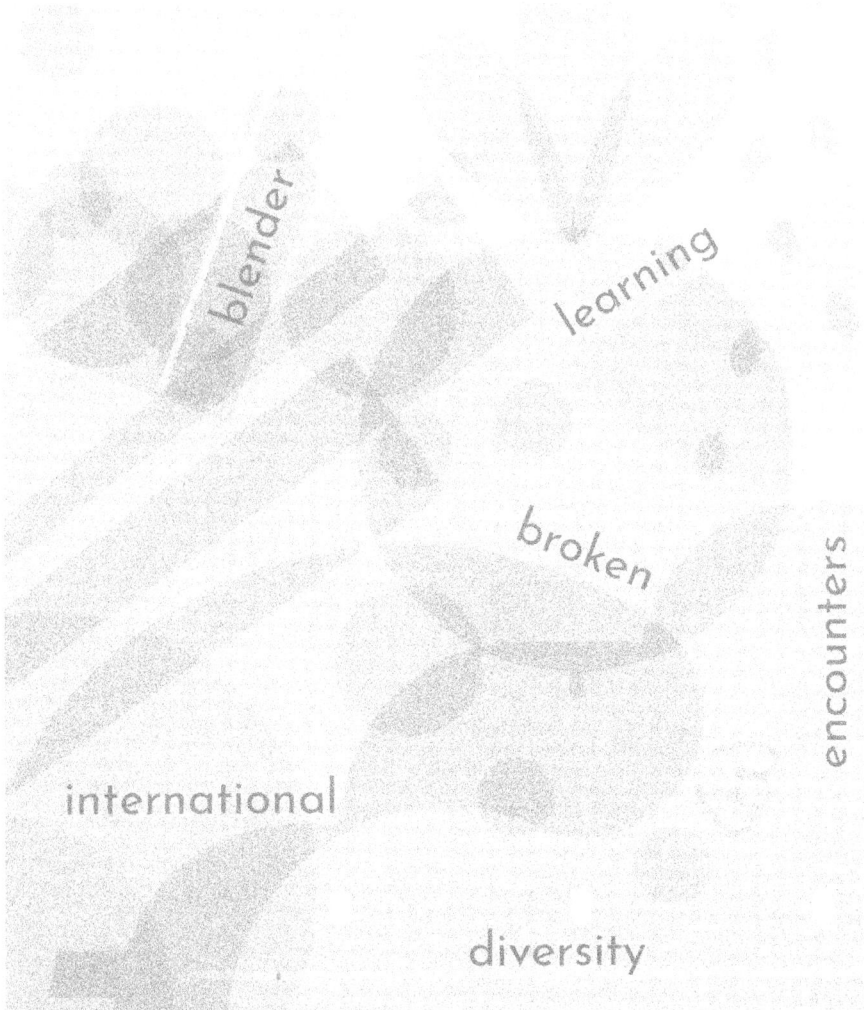

blender

learning

broken

encounters

international

diversity

6

THE INVITATION

Timing is everything

Carol Switzer

Introduction

The best idea can go unadopted if nobody knows about it. The World Challenges Programme (WCP) is no exception. This is the story of the development and implementation of the WCP at the Università della Svizzera italiana (USI). The WCP encompasses various activities and for this book, we are discussing those pertinent to the master-level course called 'Tackling World Challenges' (TWC). I was involved in the original project proposal and continue as the managing director of the project at USI. In this chapter, I will describe the marketing process that we undertook to launch to programme at USI.

The TWC started as a hopeful idea. When it got funding, things began in earnest. The funding body required that we follow a strict implementation plan with goal setting, methods to measure progress, a commitment to proceed according to the agreed upon criteria in the approved project proposal and achieve clear objectives.

Simple enough. After all, the idea originated from a marketing professor, who decided to redirect part of his discretionary budget toward consulting services (this is me, and how I came to be involved with the project). We had to define in detail where we wanted to go and how we intended to get there. In short, we had a dream scenario for any marketing professional: a great product, access to expert theory, an experienced practitioner, clear objectives and, best of all, a finite and well-known target market − the students, who were already enrolled in master's programmes at the university.

Surely students would welcome the chance to participate in such an unusual and interesting course. There was no downside. We just had to let them know about it, right? ... Easier said than done. USI has five faculties, each of which operates as a separate entity with its own regulations, separate course catalogue and different management structure. The great idea of opening up the TWC course to all faculties resulted in the not so great ramification that the course was effectively 'unlisted' – it wasn't part of any faculty list of courses offered, or any list of available courses for any programme of study. There was no place a student could go to see what courses he or she could take and find ours among them. Further, the university frowned upon the idea of promoting any one course believing, in the words of one long-time employee, 'this implied the other courses were inferior'. Hmmm.

For me, this raised another question: how does the university promote its programmes and its courses in general? USI is a Swiss public university, and, even though we offer degrees in marketing, as an institution we don't do it. Not really. This is not, in and of itself, a problem. In fact, it could be an advantage since the playing field was wide open, with none of the distortion so common in today's world where we are so heavily bombarded with advertising. On the other hand, we were not in a position to consider an overtly visible campaign directed at students when key university leadership were not inclined to support it. We knew more about our target audience than in any other marketing endeavour I've encountered, yet we were restricted from reaching out to this audience, individually or generally.

With this setback, we went back to the drawing board. What did we really want? The most basic of our goals was to enrol students in the TWC course on a competitive basis, granting entrance to one in three applicants. This meant that we needed to generate applications to the programme and to do this we needed to create visibility about the TWC – what is it and how can you get it … without traditional advertising and promotion? How could we characterise the TWC in neutral terms? How could we get key university personnel to accept (if not pro-mote) the TWC?

USI is a small place, we all know each other, and ultimately, we all want what is best for the students. Still convinced that students would apply if they knew about the course, and that TWC was good for students, we decided to tell everyone about it, in person, one by one. Effectively, we went back in marketing history, and deployed the door-to-door approach. We knew our target: all master's students. We defined an approach to recognise that this new course needed advocates, and to get them, we had to turn our colleagues into ambassadors. We wanted to ensure that the faculty and administration knew about the TWC directly from us, with all the positive ideas and energy we intended when coming up with the idea, rather than allowing them to hear about it second-hand, risking that the message become diluted, or worse, that it may be inappropriate when considering constraints students face within their pro-gramme of study. In particular, we did not want the faculty or administration to hear about the TWC from a student which might put them in the awkward and annoying position of not knowing how to respond to a student's questions.

Marketing plan

To present the marketing plan in detail, first we describe the course in typical university format, which offers a brief description and states when and where course meetings are held. Then, we state our objectives, define and analyse the market briefly, identify the marketing channels and lay out our course of action for achieving the objectives.

We came up with a brief and clear message: what TWC was, how students could apply, and by when (see Figure 6.1). This was available on the university website, outside the faculty areas, meaning that it was not in the typical path of a student searching for course options.

Our aim was to generate 36 applications from USI students, three for each available position in the course. Our target was well-defined and closed: all USI master's students could be eligible, approximately 800 individuals. To understand 'customer behaviour' (in our case, the students), we had to consider strict programme requirements that limits their access to this course, or any course outside their programme of study. Key constraints are:

- Timing of the TWC course – offered only in the spring semester.
- Travel component of the TWC course – students taking other courses or conducting an internship may not be able to travel.
- Requirements of individual academic programmes – each master's programme has a schedule that defines when elective courses can be taken. Students conducting an internship or working on their thesis may not be able to take an elective course; students taking a normal course load may have difficulty with the travel component.

The elective course, Tackling World Challenges, brings together master's students from USI, Stockholm School of Economics in Sweden and Hanken School of Economics in Finland to work in interdisciplinary and culturally diverse teams on a real-life 'wicked problem' in cooperation with an organisation.

The course consists of lectures, field trip, group work, knowledge sharing, expert interviews, company collaboration and high level presentations. There will be a kick-off meeting at the beginning of the course where students from USI will personally meet students from the other university partners at Stockholm. In the remaining seminars and lectures, students in the different schools will meet virtually and groups are expected to coordinate and connect using the available digital platform.

The course brings together 36 students (12 per university) and is structured into six interdisciplinary teams (e.g. ethics/CSR, management, communications) of six students (two per university).

Applications for Spring 20xx are open.

Places are limited so apply now to reserve your place. Course participants receive a scholarship (travel costs). To apply, send a copy of English language or other certificates you may have and a motivation piece to wcp@usi.ch.

FIGURE 6.1 Master-level course available to students from all faculties

Next, we defined benefits of the TWC course from the student's perspective, and how that translated into strengths and weaknesses for our marketing approach. In answer to the classic marketing question 'Why should they 'buy'? We give the following advantages to students:

- A modern learning environment: virtual classes, blog and discussion forum.
- Paid travel to partner universities: an international experience.
- Valuable experience: learning with students from other faculties (interdisciplinary teams), and other universities (intercultural and international experience).

This translates into several distinct advantages for students, which we could emphasise in our approach: paid travel, CV-building experience, direct contact with industry resulting in access to an international professional network, low student to professor ratio allowing high interaction with academic team.

The main weakness for reaching our target was (and still is) that there is no vehicle at USI for disseminating information about a single course or programme across all faculties. TWC is a new course. Thus, in addition to being unlisted, there was no awareness of it in the university, and it is conducted in a new format, so it would potentially have to overcome resistance to the unknown or unfamiliar. Finally, the specific timing of TWC only in the spring semester narrows the market to only students who have space in their study programme at that time. These barriers significantly reduce the market size, making our door-to-door approach even more appropriate. On the bright side, we could also prioritise certain master's programmes once we understood which ones had space for the TWC.

Before launching our marketing plan, we undertook the crucial step of gaining stakeholder buy-in. At USI, this is essentially an endorsement from the rectorate which gave us access to university services to create marketing basics: a brochure (Figure 6.1) and a web page at the coveted university-level main website, www. usi.ch/en/wcp. These basic tools reflected the backing of the university and would help legitimatise our approach to our marketing channels (master's programme managers) and our market (the students).

At this point we identified our 'market channels' and plan of action. From a tactical point of view, we opted for a pull marketing strategy rather than a push strategy since we needed to avoid actively promoting the course. The university does not promote in general, hence the promotion of one specific course would not be condoned. We needed to pull students in, to create an environment that encouraged students to apply. To do this, we first generated awareness among faculty and staff about the course, with the objective of making them ambassadors so that when students asked, faculty and staff would know how to respond at a minimum, and ideally, encourage the student to apply. We then found ways to stand in the path of students so that they would know that TWC was an option for them. To convert awareness to application, we relied on content marketing, that is, making sure that details about the class were clear and trusting that the course content would speak for itself and convince students to apply.

We defined waves of field excursions to our colleagues, first taking the time to go to administrative personnel who are often left out of communication cycles and later expected to handle related enquiries. Next, we listed all master's programmes and knocked on the door of each of the programme directors, managers and coordinators. This was not an easy task. However, by speaking face to face with our colleagues – professors and administrators alike – we were able to convey our enthusiasm for the project and, at the same time, collect concerns, constraints and suggestions from them. We did not find everyone 'at home' when we stopped by, but we followed up with an email that re-iterated the same information that we provided in person. Overall, this move, initially made for lack of an alternative, had the unanticipated consequence of strengthening the programme and smoothing its path for the future.

Implementation

Finally, we were ready to get our message to students! At this point we felt comfortable relying on the university structure to support the WCP. Happy with our old-fashioned door-to-door approach, we identified several large core classes where we could find the most students at one time, approached the professor, and asked if we could come in for the first or last five minutes of class to announce the WCP and how students could participate. Our message was not overly positive so as not to solicit comparisons to other courses; we simply informed everyone of this new option, letting them know who was eligible, how they could apply, and the relevant dates and deadlines. Significantly, we did not have any slick slide-show with images and power wording. Instead, we engaged with the students, invited their questions and, at the end, handed out our small flyer with key information about the TWC (see Figure 6.1).

All our marketing efforts were undertaken cautiously, with constant communication to university colleagues so as to ensure people were in a position to give students the information they needed to apply to the WCP. After all, we were still an unlisted class. We made sure that the next edition of the university catalogue included our course; however, there is still no place for a cross-faculty course to be seen by students unless they explicitly seek it in the overall catalogue, which is buried deep in the university database, far from where students look for course options within their master's programme.

Other marketing activities

In addition to the core activities of reaching students directly via their master's programme director and managers, we also participated in relevant university events. Due to the travel component and partnership with other universities in the course, we had a table at the International Fair, explaining to students who inquired that this was a novel and alternative way to get international experience. The TWC's short trip at the beginning of the course makes it possible for students who cannot take an entire semester to study abroad to nonetheless gain international experience. In fact, working with multinational clients in different locations with a geographically dispersed team is an intense exercise in building international acumen.

We also met with student groups, such as the university class representatives and the Association Internationale des Étudiants en Sciences Économiques et Commerciales (AIESEC) who support cross-cultural and exchange experiences for students across the globe. This was undertaken almost as a mentoring exchange exercise in the sense that they helped us spread the word about TWC and other programmes to their peers and we helped them with overview and advice on their initiatives and marketing materials.

As a gesture of support and to raise awareness for the TWC, the master's course Social Entrepreneurship undertook the subject of food waste which was the theme of the TWC as well. This course was an earlier course (offered in the autumn) and we shared research materials on food waste with them. Our newly admitted class was invited to observe the final presentations of these students at the end of their semester. The exchange of ideas was fruitful and some TWC students incorporated ideas from the Social Entrepreneurship class into their solutions.

Other 'content marketing' activities included a movie night (Figure 6.2) to raise awareness of the topic of food waste and provide details about the TWC elective course and an informative article to the student newspaper, *l'Universo* (Figure 6.3). To follow-up after the recruitment phase and generate interest for the next cycle, we ran a news item on the main page of the university website for the kick-off and final meetings of the course (Figure 6.4 and Figure 6.5).

Results

The results of our targeted marketing effort were successful. We had 30 applications for 12 positions in the class, all submitted on time and complete. We believe students respected the deadline and the requirements because they understood that course admission was competitive and that their application had to stand out among the group of applicants. There were a number of creative approaches to the motivational piece of the application, very much in line with the spirit of the class as a non-traditional, novel course – videos, blogs and presentations that described the applicant's fit with course objectives. It was a pleasure to go through the selection process. Due to the number of applicants, we were able to make a selection that was as broad as possible, admitting students with different disciplinary and cultural backgrounds. Only in gender were we unable to achieve a balance since there were few male applicants. Upon investigation, we noted that this imbalance is representative of the master's programmes of the applicants, meaning

Travel.
Study.
Tackle World Challenges.

For more info

19 Oct 2017
Int'l Relations
Info Session
12:30
Auditorium

16 Nov 2017
International Fair
12:30
Aula Magna

www.usi.ch/wcp

Contact
wcp@usi.ch

Intercultural Training
This elective course brings together Master's and PhD students from all Faculties at USI, Stockholm School of Economics in Sweden and Hanken School of Economics in Finland to work in interdisciplinary and culturally diverse teams on a real-life "wicked problem" in cooperation with an organization.

Understanding World Challenges
This year's theme will focus on food waste along the value chain: at the farm, field and factories, in the kitchen and on the plate.

Format
Lectures, field trip, knowledge sharing, expert interviews, company collaboration, and professional level training and presentations.
• Kick-off meeting in Stockholm
• Virtual meetings throughout the term
• Group work with students from partner universities

Faculty Team
• USI - Prof. Michael Gibbert, Università della Svizzera italiana
• SSE - Prof. Anna Nyberg, Stockholm School of Economics; Dr. Marijane Luistro Jonsson, Mistra Center for Sustainable Markets
• HSE Prof. Liisa Välikangas, Hanken School of Economics

Applications are open

Places are limited so get your application in soon to reserve your place. The application deadline is 1 Dec.

Università
della
Svizzera
italiana

FIGURE 6.2 Example of promotional flyer

Università
della
Svizzera
italiana

#USImovienight

Friday 6 Oct 2017 18:00
Auditorium | Free Entry

Join the class of

**Applied Social
Entrepreneurship**

for a community screening
of this award winning film

from the creators of The Clean Bin Project

JUST EAT IT.
A food waste story

Organized by **WCP**
World Challenges Program
wcp@usi.ch

FIGURE 6.3 Flyer used to promote movie night

that we had more applicants from programmes where there are more females. Our reduced target base of master's programmes that allowed flexibility for an elective course meant that certain programmes with more male students were not able to take the course due to curriculum constraints. These included finance, economics and political economy. On the other hand, the Faculty of Informatics has a strict curriculum yet very flexible timing therefore we were able to include informatics students in the TWC course.

Next steps

One of our future goals is to 'scale out' the concept of the TWC by having the university adopt the core principles of internationalism, interdisciplinary approaches and approaching wicked problems. In our meetings about the TWC, we saw a willingness among faculty to explore all of these concepts. Courses tend not to move away from traditional methods due to curriculum constraints which can be circumvented in certain cases, as with the Social Entrepreneurship class. Another constraint is faculty resources since it is effectively a new class and therefore requires the time of the professor who proposes it to teach at least a portion of the class. Building the partnership element into existing courses may be the best way to augment the use of this format without exceeding the resources available.

Another area for expansion is to carry on the themes explored in the TWC in other areas, potentially in secondary schools, adult education or the community. Through local partners, we are developing contacts and hope that this can be a long term outcome.

L'UNIVERSO
giornale studentesco universitario indipendente

☐☐L'UNIVERSO

Title
DISCOVER THE WCP WORLD CHALLENGES PROGRAMME

The WCP is a new university level initiative spanning all Faculties that aligns education, research, and engagement with key global development issues. It has grown out of the need for a common connection across emerging initiatives at USI in all Faculties, and with community based projects that address these challenges.
USI believes that attitudes and skills learned in one's early twenties are formative for later years. The integrated education part of WCP links what is learned in class to broader societal and environmental challenges through two new opportunities for students.

The first is a new inter-faculty track open to Bachelor students in the Faculties of Communication Sciences and Economics: this new track, called ICE Intercultural Communications and Economics, is in partnership with the Hanken School of Economics in Helsinki, Finland.

The program is taught entirely in English and designed for Bachelor students to increase their insights into the challenges facing modern society. Students become well balanced individuals capable of understanding the economic, societal, and environmental impact of their decisions and gain key skills that will integrate international agility with a firm grounding in the local context. Courses, workshops and other training activities put a premium on problem solving skills, intercultural competence and extramural exposure to professional work environments.

Students from USI receive a scholarship to participate in ICE and join students from Hanken School of Economics to study for a semester in Finland.

The second opportunity is an elective course for Master's and PhD students in all Faculties. Students from USI, Stockholm School of Economics in Sweden and Hanken School of Economics in Finland work in interdisciplinary and culturally diverse teams on a real-life "wicked problem". The course consists of lectures, field trip, group work, knowledge sharing, expert interviews, company collaboration, and high-level presentations.

The application deadline for both opportunities is 1 Dec 2017. To apply, complete the internal application form available at www.usi.ch/it/relint. Include the following documents with your application: most recent transcript, photocopy of language certificates and a motivation letter.

To learn more, come to the International Relations info-session on 19 Oct at 12:30 in the USI Auditorium. You can also visit the WCP table at the International Fair on 16 Nov at 12:30 in the Aula Magna.

Se possibile, inserire uno specchietto informativo
Write to wcp@usi.ch or inquire at the International Relations and Study Abroad Service Office. To apply, complete the internal application form available at usi.ch/it/relint.

Alessandro Benedetti
vice-direttore, L'universo
alessandro.benedetti@luniverso.com

FIGURE 6.4 Article by USI's student newspaper on TWC

Travel. Study. Tackle World Challenges: USI students on the move

Media and Communication Service
02/26/2018

12 USI Master's students in Economics, Informatics and Communication Sciences started the 2018 spring semester early, flying to Stockholm for a kick-off meeting of the elective course "Tackling World Challenges", offered within the USI World Challenges Programme – an inter-Faculty initiative of USI.

At the kick-off meeting in Stockholm, inter-university teams were formed among graduate-level students from USI, Stockholm School of Economics (SSE) and Hanken School of Economics (Finland). The teams will continue to collaborate over the semester, connecting virtually to work together on a project, through a common IT platform. This course represents a new internationalization model that integrates sustainable development topics into a virtual classroom environment with partner universities, giving students skills in flexible teamwork and improving their employability.

The course goes beyond traditional mobility models by not only exposing students to other European universities but giving them virtual collaboration skills to prepare them for working in a global, interconnected working world. Students form teams with peers from the Hanken and SSE, and throughout the semester the class will work with IKEA on reducing food waste in their store restaurants. Responsible consumption and waste reduction feature in the UN sustainable development goals and students will gain an understanding of how to approach complex societal issues ("wicked problems").

Read more about the World Challenges Programme. For information contact wcp@usi.ch

Sections

Education, Partnerships

FIGURE 6.5 Internal communication about TWC at USI

A further aim for the future involves creating an edition of the TWC with a university partner from a developing country. This would introduce another type of cultural exchange which we believe is valuable for students and increasingly important, as the mature European market cedes to more active developing world markets.

Reducing food waste by 50%: the ideas of 30 students

Media and Communication Service
05/10/2018

According to FAO, one third of all food produced for human consumption is wasted each year. This ethical, economic and ecological problem is further complicated by the fact that our world is divided by extreme waste on the one hand, and victims of famine on the other. On Wednesday 9 May, a group of 30 students from USI, the Stockholm School of Economics (Sweden), and the Hanken School of Economics (Finland), gathered in Lugano to present various ideas to contribute creatively and effectively to the fight against food waste.

The students, all participants in an innovative international inter-faculty course offered as part of the USI World Challenges Program, developed their proposals utilizing data from IKEA restaurants in Lugano, Stockholm and Helsinki. The official collaboration with the Swedish multinational enabled students to come up with concrete ideas on how to reduce the 50% a further 50% by 2020.

These ideas are not just intended to involve customers in the act of purchase and consumption in restaurants, but also aim to increase sensitivity on the topic at home, in the course of everyday life. Ideas range from social media campaigns to sustainable recipe books, from "museums" on waste to actual signs inside shops, promoting – in accordance with the principles of behavioral economics - a more responsible way of acting.

After the presentations, students moved to the kitchen for a "Master Chef" style competition - coordinated by Gaetano Tresoldi - focusing on optimizing ingredients as well as taste: this part of the initiative is among the affiliated activities of Lugano città del gusto (city of taste) 2018, which, like the USI World Challenges Program, aims to promote a culture more attentive to the complexity of the food sector, especially among young people. Lugano Mayor Marco Borradori also took part in the competition, bringing greetings from the City Hall to our visitors from Sweden and Finland.

The course, that began with USI students traveling to Stockholm and involves active collaboration and exchanges between USI, the Stockholm School of Economics and the Hanken School of Economics is supported by Mercator Swiss. The meeting in Lugano is supported by the Swiss Academy of Humanities and Social Sciences (SAGW).

For more information: https://www.usi.ch/en/wcp

Faculties

Faculty of Communication Sciences

Sections

Education, Partnerships, Campus Life

FIGURE 6.6 USI's communication services runs article on TWC

We believe the model of the TWC course can benefit teachers as well as students, enhancing the opportunity for knowledge transfer back into research from classroom activities. Considering the major difference the TWC approach offers to traditional methods, it is relatively straightforward to implement and requires little investment if applied to existing curricula. The key is to have strong support from university leadership, and a clear plan for which areas of the curriculum are good candidates for incorporating aspects of the TWC.

7

A LEARNING BLENDER

How virtual team set-up influences outcomes

Monika Maślikowska

Introduction

Although the concept of working in virtual teams is not new (Jarvenpaa & Leidner, 1999; Orlikowski, 2000), what we know about the effectiveness of virtual teams in management education still leaves a lot to be desired (Davison, Panteli, Hardin & Fuller, 2017; Taras et al., 2013). Setting up Global Virtual Student Teams (Davison et al., 2017) and guiding them to their final output and performance still require intensive exploration. This is important, because GVSTs are different from their business practice equivalents since they lack a power structure and have a relatively flat hierarchy, typical for student teams (Davison et al., 2017), but also due to different motivation and external expectations towards outcomes. While for professionals, the primary outcome of work and performance might be a practical application of work and compensation as a reward, for students the main output is instead learning and receiving feedback. The approach to the topic becomes even more complicated in the growing context of sustainability mixed into business school curricula. For this reason, educators would be wise to take into account the associated complexities of increased diversity and multidisciplinarity (Hardy & Tolhurst, 2014) while tackling wicked problems (Ferraro, Etzion & Gehman, 2015; Rittel & Webber, 1973). Furthermore, academia needs to adapt to new forms of intelligent working and virtual collaboration (Bradley & Woodling, 2000; Maślikowska & Gibbert, 2019; Taras et al., 2013; Vischer, 2008; Wei, Thurasamy & Popa, 2018), which is particularly important with the increasing trend towards meta-teams (Santistevan & Josserand, 2019), fluidity, overlap and dispersion in organisation studies (Mortensen & Haas, 2018).

The World Economic Forum (2016) reports that young people are crucial drivers in approaching the Sustainable Development Goals (SDGs); however, they need to obtain the necessary skills and opportunities, such as interdisciplinary experience and education, and international collaboration (Sustainable Development Goal 17—United Nations University, n.d.; UNESCO Digital Library, n.d.). In addition, what we observe in studies on teaching sustainability within business education is that teaching students to be integrated catalysts (Akrivou & Bradbury-Huang, 2015) is essential in approaching sustainability challenges. In order to address these issues in the TWC course, we decided to follow the concept of *bricolage* (see Chapter 12) and the *phronetic* approach to management education where business practitioners and academics are co-participating (Billett, 2002) in the creation of knowledge (Antonacopoulou & Bento, 2018; Antonacopoulou, 2010). This approach not only allows students to obtain practical knowledge and leadership skills (Antonacopoulou & Bento, 2018) while working in GVSTs in the applied context but also brings benefits to all parties involved in the project through the integration of teaching (learning outcomes for students and everyone involved!), research (academics) and community engagement (practical solutions for business partners). As the World Challenges Programme coordinator and a PhD student, I was responsible for the administrative and organisational side of the TWC course (e.g. student recruitment, organisation of the final summit), as well as an academic part (contribution to teaching and research). From my position between the students and professors, and academia and practice, I had a unique opportunity to lead and observe different sides of the project and have access to first-hand information from different parties involved: business partners, professors and assistants, administrative staff and students. In the following sections of this chapter, I will share my experience and observations on setting up our Global Virtual Student Teams, starting from student recruitment, selection and grouping based on diversity, through designing an environment for increased performance and finally including the student project outcomes presented during the final summit in Lugano, Switzerland (2018).

Student-selection process and grouping based on diversity

While promoting the course during the fall semester (see Chapter 6 in this book), we wanted to make sure that we attracted students interested in gaining practical international experience working for an industry partner in interdisciplinary virtual teams and learning while working on the global challenges of sustainability and food waste. At the same time, we wanted to select highly motivated candidates who could benefit the most from the course and contribute to their best ability. As we designed the course to function across universities, disciplines and faculties in interdisciplinary (Huutoniemi, Klein, Bruun & Hukkinen, 2010) teams working on wicked problems, we recruited master's students from all faculties in all three partner universities: Stockholm School of Economics (SSE), Hanken School of Economics (Hanken) and Università della Svizzera italiana (USI). Students were assigned to these interdisciplinary, culturally diverse, and internationally distributed Global Virtual Student Teams (GVSTs), which worked together on global tasks across space, time and organisational boundaries, conducting most of their interaction and decision-making using communication technology (Davison et al., 2017; Maznevski & Chudoba, 2000). For the first year of the course, class participants worked on implementing creative ideas via project development and physical interventions, contributing to IKEA's goal of reducing food waste by 50% globally in their restaurants by 2020. The course agenda addressed several of the United Nations Sustainable Development Goals (2030 Agenda), primarily the 'Responsible Production and Consumption', but also 'Zero Hunger', 'Quality Education' and 'Partnership for the Goals' ('Sustainable Development Knowledge Platform', n.d.) to explore various approaches to resolving wicked problems (Rittel & Webber, 1973).

Course capacity and the funding agreement allowed for enrolling 12 students from USI, so for the sake of a fair balance we aimed to launch the course with 12 students per university, in total 36 students. We set a target of recruiting 36 applicants for the 12 available positions at USI so that we could select students who most closely matched our requirements. To ensure that the main selection criteria were met, we evaluated applicants based on 'motivation pieces', academic transcripts with Grade Point Averages (GPA) and language skills. Our primary focus was the motivation pieces, which ranged from traditional letters to blogs, videos, presentations and websites. We focused on the content, subject understanding, international and practical experience or a demonstrated interest (we also wanted to be open to less experienced students who could gain these skills during the course), extramural activity, quality of communication and presentation (creativity, originality, design, composition, appropriate length, structure, clarity; was it funny, interesting, inspiring, provoking, unconventional, mind blowing?) and reasons why the student would like to pursue the course. In order to guide students through the entire process and gain up-to-date information about the course and application procedure, we created a blog, where we posted application tips, for example, 'Tell us something interesting about you, describe your background and explain your decision and motivation for applying to the programme. Why do you feel you are a strong candidate for the programme? How will this programme influence your career intentions? Be creative!'.

We also considered the diversity of the overall class in terms of student backgrounds (study focus or major) and individual characteristics including attitude, giving special attention to proactive and highly motivated students (experience, interests). The main purpose of the academic transcript and GPA was to verify that students were on track with their studies (and therefore eligible) and note their performance. Since we are aware that grades might not always be representative of a candidate's entire potential and the candidate motivation and diversity was our primary focus anyway, we decided to waive this part of the application in the future. The English language skills, on the other hand, continue to be required to assure that in the already complex configuration of the course, at least communication in a common language is assured.

We assumed that the complementary diversity of the highly motivated members of the Global Virtual Student Teams would be the main tool and strength in approaching wicked problems, so once we selected our top applicants and confirmed their enrolment, we approached the grouping. In the course edition 2018 we managed to enrol 29 students in total (6 from SSE, 11 from Hanken, and 12 out of 29 who applied from USI), so we decided to divide them into five optimal size groups of 5–6 students (Hackman, 2002). The diversity of the selected candidates included universities (3), education (e.g. MSc in Cognitive Psychology, Management, International Tourism, Informatics, Public Management and Policy, Financial Communication), nationalities (15: Austrian, Bulgarian, Finnish, French and Swedish, Greek Cypriot, Italian, Luxembourgish, Spanish, Swedish, Swiss, Tunisian, American, German, Chinese), gender (7 male, 22 female) and age (from 22 to 37). In order to maximise the diversity in each team, we made sure that each group had a balanced number of students from each university, study stream, country (5–6 per group), gender (at least one male per group and we tried to increase that number in the next course editions) and age (mixed ages in each group). In the next step, course participants (GVSTs) and staff (professors, lecturers, teaching assistants, IT support and trainers) met in the face-to-face kick-off meeting in Stockholm, Sweden.

Designing a high performance work environment

The course schedule designed for the spring academic semester included:

- a face-to-face kick-off meeting hosted by one of the Scandinavian partner universities (February);
- a series of virtual encounters composed of lectures and consultations (led by the professors and assistants of the partner universities, as well as guest speakers);
- individual meetings of the virtual student teams;
- a final face-to-face meeting in a form of a final summit in Lugano, Switzerland (May).

In the following latest course generations, despite the relatively large proportion of the virtual collaboration (vs. face-to-face), we decided to maintain the highly beneficial initial face-to-face meeting (both for the students and the project itself), but cancel the final meeting. The goal was to keep the most impactful and crucial elements of the course, but reduce the number of flights, which in turn reduces our carbon footprint and reflects our desire to be more environmentally sustainable.

The kick-off meeting includes lectures (from guest speakers, professors and assistants and external partners), training (IT on digital tools and external organisations on food waste) and activities (team-building, informal dinner, brainstorming and creativity sessions), providing an intensive start to the virtual collaboration and professional networking. One of the main purposes of the kick-off meeting has been to set the stage for increased performance from the Global Virtual Student Teams which, according to several researchers, can be supported by team cohesion (Hoegl & Gemuenden, 2001; Hoegl, Gibbert & Mazursky, 2008), and associated

Evaluation Breakdown

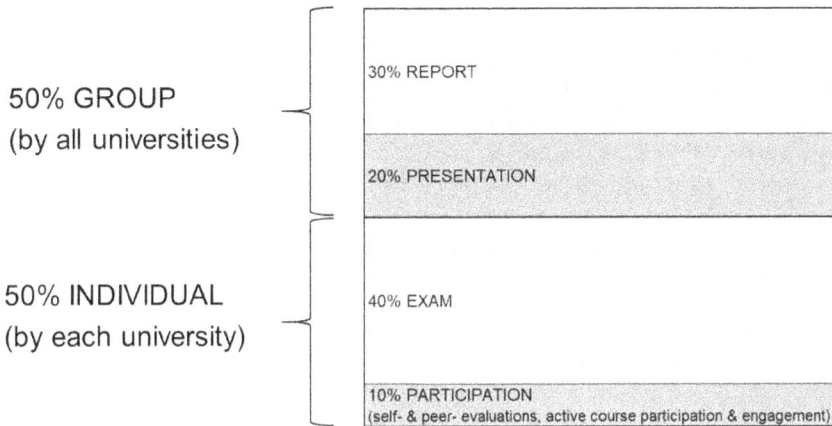

50% GROUP (by all universities)	30% REPORT
	20% PRESENTATION
50% INDIVIDUAL (by each university)	40% EXAM
	10% PARTICIPATION (self- & peer- evaluations, active course participation & engagement)

FIGURE 7.1 Evaluation breakdown

motivation (Hertel, Geister & Konradt, 2005) and trust (Paul, Drake & Liang, 2016). We aim to assure high motivation through appropriate student-selection based on diversity and inclusion and the development of a motivating grading system composed of 50% individual performance and 50% team performance; individual evaluation consists of an exam (literature review and personal reflection essay) and peer- and self-assessment, while team evaluation is based on the final presentation and report (Hertel et al., 2005). Figure 7.1 shows the course evaluation breakdown.

Trust building, on the other hand, was initiated through the face-to-face kick-off meeting, which clarified team goals (Hertel et al., 2005) and offered team-building activities (Chang, Hung & Hsieh, 2014).

During the semester and after the kick-off, remote collaboration between student teams, professors, external partners and guests occurred mostly in virtual spaces using several technological means, including Microsoft Teams (principle team management tool, for which training was provided), Google Drive, Skype, WhatsApp, Facebook, email, Moodle. At the same time, all course participants had the flexibility to choose their physical working space and were given very few guidelines (as opposed to restrictions and requirements) on how their work should be conducted. Such an approach allowed each team to naturally develop their own dynamic, working styles and structure, which influenced their collaboration and in consequence had a major impact on the overall team (and individual) performance. Apart from the video conferences (lectures, consultations, team meetings), calls, emails, texts, recorded messages, etc., students also had the opportunity to meet in person on the national level to implement their intervention ideas at the IKEA stores in three countries (Switzerland, Sweden, Finland).

Surprisingly, after a semester of collaboration, we observed the unusual symptoms of 'dynamic boundaries' among the Global Virtual Student Teams participating in the project. Students working on wicked problems were not only working in physical dispersion. They were also participating in several team projects at the same time, moving on and off teams (swapping roles with other team members), formulating sub-teams, collaborating across teams, and while often ignoring the concept of traditional team cohesion, they intended to increase performance in addressing the wicked problems of food waste and sustainability. This trend, aligned with the modern tendency of group collaboration to move towards 'dynamic participation hubs' (Mortensen & Haas, 2018) seems to have brought positive outcomes in the form of very creative final student projects.

Final student projects

On 9 May 2018, TWC students from the three partner universities met in person to present the results of their food waste reduction projects for IKEA with an event located in Lugano, featuring the USI Rector, Mayor of Lugano, international professors and local organisations involved in the gastronomy business. The interdisciplinary and diverse student teams of the Tackling World Challenges cross-university virtual course provided creative ideas and developed projects that contributed to IKEA's goal of halving food waste in their restaurants globally by 2020. The solutions were applied at the consumer level and the purpose of the interventions was to not only to reduce food waste within the IKEA restaurants but also to increase awareness on the topic and influence customer behaviour outside the stores and at home.

Students collaborated internationally (virtually) to collect data on food waste through surveys, observations and interviews with store customers. Based on their findings, they applied targeted solutions in the IKEA stores in Lugano, Stockholm and Helsinki. Following a stage of brainstorming, students provided multiple ideas, after which, with the consultation of the corporate partner and professors from each university, they selected and developed one project per team. Projects ranged from the Instagram Campaign #ILCIBOÈPREZIOSO (#FOODISPRECIOUS) inviting customers to the food-waste photo competition, to THE FOOD WASTE MUSEUM and FOOD WASTE DRAWING CAMPAIGN – collaborative initiatives helping to spread awareness on the topic of food waste among families with kids, who were a main target of the campaigns due to the high impact level. Other ideas developed involved NUDGING SIGNS displayed in prominent places of the IKEA restaurants or SUSTAINABLE RECIPES distributed to customers.

Food-waste projects at glance – summaries provided by student teams

#ILCIBOÈPREZIOSO

Our idea was to introduce social media as a communication channel against food waste. We chose to use Instagram as our social media channel as it is a photo-oriented platform. As IKEA Grancia had no Instagram account up to that point, we created a new one dedicated specifically to tackling food waste. After building the Instagram account, #ILCIBOÈPREZIOSO hashtag was created to make the campaign viral and gain popularity and followers. Our idea was to encourage IKEA customers to consume everything on their plates, take a photo of the empty plates or take a selfie with it, and post it on IKEA Grancia's Instagram account.

To provide an incentive for customers to participate in the Instagram food waste campaign, a "winner of the day" was selected randomly to receive a reward. See Figure 7.2 for an example of a post from the campaign. Besides the reduction of the overall food waste in the restaurants, the idea contributes to the general marketing portfolio of the IKEA brand and positioning IKEA as a food waste champion. Moreover, this type of campaign has an ease of replicability across stores, and the potential to reach many people, even those who do not visit the restaurant often. (Student report)

THE FOOD WASTE MUSEUM

The final proposal selected by IKEA's *Project Leader – 'IKEA Food Waste Watcher Programme'* (Signe Damgaard Nielsen) was to create an *IKEA Food Waste Museum*, where food waste would be explored with all senses. However, since the implementation time for this project was only one-week after

FIGURE 7.2 Example of student project

communication/verification and certain resource-related constraints existed, the scope of the implementation was scaled down. Therefore, an *IKEA Food Waste Event* was organised. The primary goal was to educate IKEA customers about food waste in an innovative and fun way, focusing on a target segment of young children aged 2 to 16, while not excluding parents and adults, and channelling everyone's inner child. Figure 7.3 shows images from the event. The narrow focus on children came from a specific idea; currently uninformed or less informed, children will soon become the adults of this world and if they are raised in a culture of not wasting then hopefully they will hold this as a core value and continue to pass down the information and solutions to food waste to future generations to follow. Thus, the solution to food waste will be approached differently in the future compared to today – we hope. (Student report)

FOOD WASTE DRAWING CAMPAIGN 'Leading by Example'

Solution *Leading by Example* was chosen by IKEA as most feasible and with the highest potential of being successfully implemented within this short time frame. In our team, we aimed to: 1) implement something innovative, different than usual communication tools, and 2) actively engage children and parents, in a fun and positive way. For this reason, the proposed solution *Leading by Example* was modified and improved in order to mix both direct and indirect engagements. This was also aligned with our two main observations – namely that 1) families with children tend to waste more food than other customers, and 2) even though people are aware of the food waste issue, they still waste food at IKEA. This could be explained by the fact that customers primarily perceive IKEA as a furniture store, and do not perceive IKEA's restaurants similarly to other fine eateries. Some customers may negatively perceive the value of IKEA menus being produced at large-scale. We believed that by combining traditional nudges (table talkers and posters aiming to raise awareness, promote IKEA's effort and set the standards) with our active and fun *creative drawing wall*, our solution could be original but efficient and impactful. See the drawing walls and parent child engagement in Figure 7.4. The next part of the report will provide an in-depth analysis of its implementation. (Student report)

NUDGING SIGNS

One of the stated goals of this project is to prevent food waste at IKEA, while avoiding intimidating customers who leave food on their plate. If reductions in food waste are seen as intrusive or authoritarian, this may lead to resistance from customers (Wansink, 2007). Therefore, nudges are an optimal solution as they are not perceived as reducing or threatening a customer's sovereignty (Just & Hanks, 2015). A nudge is defined as 'any aspect of the choice architecture that alters

EVENT DAY REALIZATION

W H A T   R E A L L Y   H A P P E N E D ?

RIGHT PLACE AT THE RIGHT TIME

HIGH RESPONSE

CURIOSITY

ITALIANO, ITALIANO, ITALIANO...

MOTIVATED FOR COLORING

MISSING GROUP MEMBERS

TIME SPAN

CHILDREN, FAMILY, AND ACTIVE PARTICIPATION

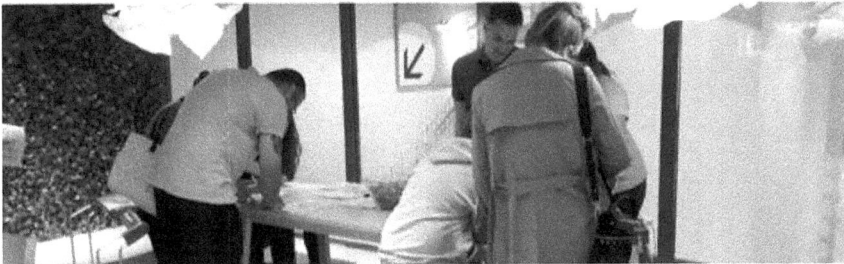

FIGURE 7.3 IKEA food waste event

FIGURE 7.4 Creative drawing wall

people's behavior in a predictable way without forbidding any options or significantly changing their economic incentives' (Thaler & Sunstein, 2008, p. 6). Research suggests that people tend to be extremely sensitive to perceived social norms. We are biologically hardwired to follow the crowd (Just & Swigert, 2016). Thus, nudges that convey social norms are extremely effective. One prominent example is the study by Kallbekken and Saelen (2013) from the CICERO Center for International Climate and Environmental Research in cooperation with Nordic Choice Hotels, the biggest hotel chain in the Nordic countries. They show that two simple and nonintrusive nudges – reducing the plate size and providing social cues – reduced the amount of food waste in hotel restaurants by around 20% with no influence on guest satisfaction. More specifically, they displayed a sign at the buffet stating 'Welcome back! Again! And again! Visit our buffet many times. That's better than taking a lot once', thus making it salient for restaurant guests that it is socially acceptable to help themselves more than once from the buffet. Just introducing the sign reduced the food waste by 20.5% (p<0.0001).

We implemented one of the ideas, namely putting up nudging signs at the IKEA Kungens Kurva breakfast buffet from 09.30 to 11.00am, at two different places in the restaurant. Figure 7.5 shows the sign and locations. Both signs depicted the same text, in both English and Swedish: 'Welcome back! Again and Again! Visit our buffet many times. That's better than taking a lot once. (Student report)

SUSTAINABLE RECIPES

The idea 'sustainable recipes' is to compose, design, produce and finally sell recipe books that contain sustainable recipe ideas and hints that help to reduce food waste in everyday life. Included recipes are sustainable in that they creatively use parts of foods that are otherwise wasted, items close to deterioration or simply facilitate food management at home. There are a number of benefits that can be associated with the idea. It is very easy to implement, as it does not require a huge upfront cost to design and print recipes. Furthermore, it does not directly interfere with daily business at both IKEA and IKEA restaurants, i. e. there is no additional training required for employees or additional staff requirements. Besides, most visitors of IKEA and its restaurants are young couples and families, who more frequently cook at home than any other age groups. As such, recipe books are directly associated with their needs and have a stronger impact. Unlike ideas which aim to have a direct impact on reducing

Nudging sign

On the food counter

At the entrance

- Nudging sign states "Welcome back! Again and Again! Visit our buffet many times. That's better than taking a lot once" in **English and Swedish**

- Presented at **two different places** in the restaurant

- Displayed **every day** during buffet hours from **09:30 to 11am**

FIGURE 7.5 Nudging sign examples

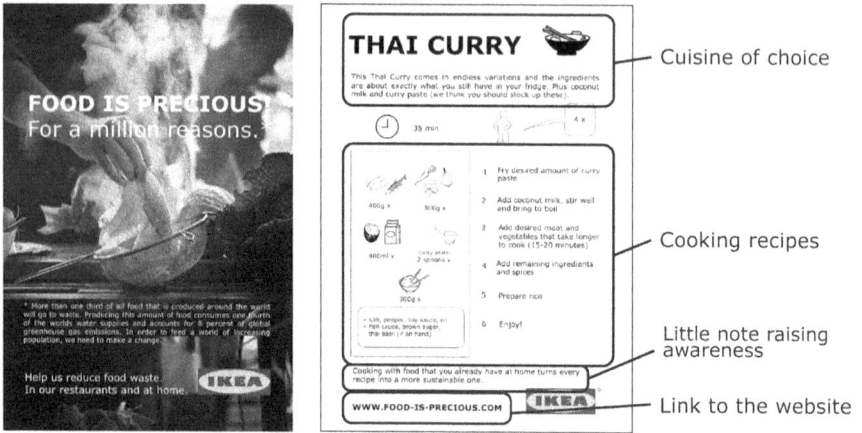

FIGURE 7.6 Example of student project

food waste at IKEA restaurants, recipe books are meant to be taken home. Our aim is to invite customers to engage with the topic of food waste and spark thoughts, discussions and of course action. As customers start connecting the IKEA brand with sustainability, we expect a change in behavioural patterns at the restaurants. Figure 7.6 shows our sustainable recipe flyer.

With this idea we hope to provide a new perspective on IKEA's potential to tackle food waste, both at their restaurants and at home, and to accordingly raise ambitions. We believe that a recipe book, given they are well designed, can have long lasting impact on the campaign against food waste. (Student report)

References

Akrivou, K. & Bradbury-Huang, H. (2015). Educating integrated catalysts: Transforming business schools toward ethics and sustainability. *Academy of Management Learning & Education*, 14(2), 222–240. https://doi.org/10.5465/amle.2012.0343

Antonacopoulou, E.P. (2010). Making the business school more 'critical': Reflexive critique based on phronesis as a foundation for impact. *British Journal of Management*, 21, s6–s25. https://doi.org/10.1111/j.1467-8551.2009.00679.x

Antonacopoulou, E.P. & Bento, R.F. (2018). From laurels to learners: Leadership with virtue. *Journal of Management Development*, 37(8), 624–633. https://doi.org/10.1108/JMD-12-2016-0269

Billett, S. (2002). Workplace pedagogic practices: Co-participation and learning. *British Journal of Educational Studies*, 50(4), 457–481. Retrieved from JSTOR.

Bradley, S. & Woodling, G. (2000). Accommodating future business intelligence: New work-space and work-time challenges for management and design. *Facilities*, 18(3/4), 162–167. https://doi.org/10.1108/02632770010315733

Chang, H.H., Hung, C.-J. & Hsieh, H.-W. (2014). Virtual teams: Cultural adaptation, communication quality, and interpersonal trust. *Total Quality Management & Business Excellence*, 25(11–12), 1318–1335. https://doi.org/10.1080/14783363.2012.704274

Davison, R.M., Panteli, N., Hardin, A.M. & Fuller, M.A. (2017). Establishing effective global virtual student teams. *IEEE Transactions on Professional Communication*, 60(3), 317–329. https://doi.org/10.1109/TPC.2017.2702038

Ferraro, F., Etzion, D. & Gehman, J. (2015). Tackling grand challenges pragmatically: Robust action revisited. *Organization Studies*, 36(3), 363–390. https://doi.org/10.1177/0170840614563742

Hackman, J.R. (2002). *Leading teams: Setting the stage for great performances.* Boston, MA: Harvard Business School Press.

Hardy, C. & Tolhurst, D. (2014). Epistemological beliefs and cultural diversity matters in management education and learning: A critical review and future directions. *Academy of Management Learning & Education*, 13(2), 265–289. https://doi.org/10.5465/amle.2012.0063

Hertel, G., Geister, S. & Konradt, U. (2005). Managing virtual teams: A review of current empirical research. *Human Resource Management Review*, 15(1), 69–95. https://doi.org/10.1016/j.hrmr.2005.01.002

Hoegl, M. & Gemuenden, H.G. (2001). Teamwork quality and the success of innovative projects: A theoretical concept and empirical evidence. *Organization Science*, 12(4), 435–449.

Hoegl, M., Gibbert, M. & Mazursky, D. (2008). Financial constraints in innovation projects: When is less more? *Research Policy*, 37(8), 1382–1391. https://doi.org/10.1016/j.respol.2008.04.018

Huutoniemi, K., Klein, J.T., Bruun, H. & Hukkinen, J. (2010). Analyzing interdisciplinarity: Typology and indicators. *Research Policy*, 39(1), 79–88. https://doi.org/10.1016/j.respol.2009.09.011

Jarvenpaa, S.L. & Leidner, D.E. (1999). Communication and trust in global virtual teams. *Organization Science*, 10(6), 791–815.

Just, D.R. & Hanks, A.S. (2015). The hidden cost of regulation: Emotional responses to command and control. *American Journal of Agricultural Economics*, 97 (4), 1–15.

Just, D. & Swigert, J. (2016). The role of nudges in reducing food waste. In *Food security in a food abundant world* (pp. 215–224). Published online. Retrieved from https://doi.org/10.1108/S1574-871520150000016009

Kallbekken, S. & Saelen, H. (2013). Nudging hotel guests to reduce food waste as a win-win environmental measure. *Economic Letters*, 119, 325–327.

Maślikowska, M. & Gibbert, M. (2019). The relationship between working spaces and organizational cultures. *Facilities*. https://doi.org/10.1108/F-06-2018-0072

Maznevski, M.L. & Chudoba, K.M. (2000). Bridging space over time: Global virtual team dynamics and effectiveness. *Organization Science*, 11(5), 473–492.

Mortensen, M. & Haas, M.R. (2018). Perspective—Rethinking teams: From bounded membership to dynamic participation. *Organization Science*, 29(2), 341–355. https://doi.org/10.1287/orsc.2017.1198

Orlikowski, W.J. (2000). Using technology and constituting structures: A practice lens for studying technology in organizations. *Organization Science*, 11(4), 404–428.

Paul, R., Drake, J.R. & Liang, H. (2016). Global virtual team performance: The effect of coordination effectiveness, trust, and team cohesion. *IEEE Transactions on Professional Communication*, 59(3), 186–202. https://doi.org/10.1109/TPC.2016.2583319

Rittel, H.W.J. & Webber, M.M. (1973). Dilemmas in a general theory of planning. *Policy Sciences*, 4(2), 155–169.

Santistevan, D. & Josserand, E. (2019). Meta-teams: Getting global work done in MNEs. *Journal of Management*, 45(2), 510–539. https://doi.org/10.1177/0149206318793184

Sustainable Development Goal 17—United Nations University. (n.d.). Retrieved from https://unu.edu/explore/sustainable-development-goal-17

Sustainable Development Knowledge Platform. (n.d.). Home. Retrieved from https://sustainabledevelopment.un.org/

Taras, V., Caprar, D.V., Rottig, D., Sarala, R.M., Zakaria, N., Zhao, F ... & Zengyu Huang, V. (2013). A global classroom? Evaluating the effectiveness of global virtual collaboration as a teaching tool in management education. *Academy of Management Learning & Education*, 12(3), 414–435. https://doi.org/10.5465/amle.2012.0195

Thaler, R.H. & Sunstein, C.R. (2008). *Nudge: Improving decisions about health, wealth, and happiness*. New Haven, CT: Yale University Press.

UNESCO Digital Library. (n.d.). Retrieved from https://unesdoc.unesco.org/ark:/48223/pf0000245656

Vischer, J.C. (2008). Towards an environmental psychology of workspace: How people are affected by environments for work. *Architectural Science Review*, 51(2), 97–108. https://doi.org/10.3763/asre.2008.5114

Wansink, B. (2007). *Mindless eating: Why we eat more than we think*. New York: Bantam.

Wei, L.H., Thurasamy, R. & Popa, S. (2018). Managing virtual teams for open innovation in global business services industry. *Management Decision*, 56(6), 1285–1305. https://doi.org/10.1108/MD-08-2017-0766

World Economic Forum. (2016). *Report*. Retrieved from http://www3.weforum.org/docs/WEF_Future_of_Jobs.pdf

8

VIRTUAL ETIQUETTE

Encounters among digital natives

Gottfried Gemzell

Introduction

The word 'consequences' can be used in both a negative and positive context. Those of us who deal with the consequences of digitalisation know this only too well. It's nice to work anywhere and connect to the world with a small computer using the Internet. In this situation, the consequences are that people feel free, can access almost any information they need in a heartbeat and collaborate with a group consisting of individuals that could be spread out all over the world.

Other consequences are that public study areas at universities around the world are getting crowded, since it's nicer to sit in a beautiful campus surrounded by friends to write and collaborate than in your one square centimetre student flat with dishes piled up and dirty laundry. Those of us born just a bit too long ago didn't have the luxury of choice since it was not possible to lug a typewriter or an old computer around on public transportation. We were stuck in that room to write an essay, typewriter sounding like a machine gun, neighbour banging a broomstick on the ceiling, because one person's floor is always someone else's roof, until we all just gave up and went to the pub (some things will stay the same through the decades).

Imagine that: a demand for more physical study spaces on campus is often a consequence of digitalisation. This implies that we have to take a broader view on digitalisation than just looking at systems, software, new machinery, and really open up the scope to understand the new digital world with its very strange natives (that is, of course, if you are old like me).

Let's briefly touch on some digitalisation developments we all know have affected higher education in recent years, just to warm up and get a glimpse of the context of it all.

Why even bother going to the lecture?

The content of the lecture is most likely retrievable online today. In a globalised world, it's even a possibility that the authors of the required reading or creator of the model has made their own lecture available on the Internet. Or that the best expert, a brilliant speaker or teacher, is holding a webinar open to everyone, recorded and uploaded for free on a website easily accessed by your students. Now, that's a depressing thought: you who enjoyed running are now competing with Usain Bolt.

The answer can be to flip the classroom. Ask the students to watch and read all the material before the lecture, and then use the lecture to discuss, clarify and analyse the subject. Watch Usain Bolt run and comment on his untied shoelace.

What is a degree worth if anyone can take an online course offered by the best universities in the world? University degrees are in a state of hyperinflation. If everyone has a master's degree, then it means nothing anymore. The thrill is gone. Or is it?

The reason why the TWC class has been a topic of conversation here on campus is because it is a course taught with the latest of digital means, with a lot of online tools, but at the same time it's a very classroom-bound course, with three professors from three different well-known universities taking turns lecturing. The lectures are intense, with three classrooms in three countries, following the lecture in real-time. Between the lectures, the students work in groups to complete a project that runs for the entire course. Every student group consists of members from all three partner universities, and the students are the next masters of the globalised world with the simple assignment of getting us out of this mess of wicked problems we find ourselves in. So surely, I'd say that this is a pursuit that counts. But how can we help them collaborate in these transnational groups so that they can save our planet later on?

After more than 12 years as a digital pedagogue, part-time and full-time, I have made some discoveries regarding team collaboration in a digital environment. Of course, the means of collaboration have changed and improved a lot, but some of the key challenges remain. We are, after all, monkeys with no tail.

Group work challenges

As a rule, we are more likely to *feel* like we are working better with people in our close physical proximity. And in some cases, we do. In a group consisting of members connecting from a distance to the group discussion or contributing to a joint effort to produce a document, the people working together in a common physical space are more often in charge of the group. The person connecting from a distance often feels left out or more contrary to a joint agreement already reached by a group in the same physical space.

There are of course always exceptions. However, we have learned that addressing this issue with the students, making them aware of this possibility, increases

their empathy and awareness to include the distant members more in group decisions and other collaborative activities. This is something we address with the students in the kick-off meeting for the World in The Making course.

Another thing we stress is that comments written down and sent to group
members by email or in a digital forum from people in the group tend to be perceived as harsher than if those same words would have been spoken by a person in
the same room. If you have bad news, don't email! Try instead to call the person
on the phone or engage in a virtual meeting. Looking somebody in the eye makes
it easier than you think to be eye to eye with somebody.

Factors that affect group work in a digital environment

Time zone, language, quality of connecting to the Internet, are all things that affect
a group connecting in to become a team. I've personally been in teams with
members around the world that outperformed a team located in close geographic
proximity. The secret behind our success was zero sleep. There was always one
team member awake, working under a never-setting sun with problems that came
from a place that never seemed to catch any of that light.

Gather around, digital natives, and let me tell you a story: the first time I collaborated on a digital platform was with two other team members from three different geographical locations on a work assignment. (I'm too old, remember, to
have had this experience as a student.) This is from my days as a consultant in the
research and development field working with international university research
centres. It was the first time around for everybody and it was very exotic, especially
working for a department of the Swedish government.

I got up in the morning and checked the team platform for the night's progress. My colleague in Singapore, Thomas, whom I've never met in person and
had a picture of Nixon as an avatar on the platform, was a workaholic like
myself. During my (too) short sleep, he had completed a lot more tasks than I
had done the previous night for him, day for me. (Or was it the other way
around?) Anyway, I was stressed out. Even my avatar reflected this – a photo of
Bob Lazar looked red-eyed and completely drooling out of sleep deprivation.
The third team member, Debbie, living in Texas, had an avatar of an old woman
in an obscene pose on a mountaintop. She was smarter than Thomas and me and
did some kind of audit and proofreading of our work. We were documenting
interviews, analysing different scenarios and compiling everything into a big
report that meant life or death for high tech projects in international research
centres.

It was very futuristic for a person who had been a teenager in the 1980s to get
up, stare at the computer screen for ten minutes for a quick review of Thomas'
work. Drinking coffee, making breakfast for the entire family, yelling at everyone
to please hurry. Remembering the picnic-day at kindergarten, flipping pancakes at
the last second while listening to mp3 files of interviews with researchers explaining the latest thing in nano-robotics. This was in 2007 and I told everyone who

was close enough to listen that audio files were the new hype. The absolute best of the best. 'I can fry pancakes and read a book or listen to the most boring interview in the history of the world.' No one would believe me for years.

It was a nice thing to work solo until Debbie came on the platform at four in the afternoon. At that time, I would have a quick beer in the pub on Grevgatan in Stockholm before going to kindergarten to pick up my son. I saw her message on my brand new iPhone. The previous year I had worked for a company desperately trying to sell a navigation software for mobile phones. Everybody speculated on the future boom of smartphones. A year later we had underachieved on our wildest dreams. 'How was your day?' she would say in a nice message, continuing not so nice 'you will have to do a retake on the interviews with the research centre ALUPO tomorrow. The one you did was absolutely crap.' Thank god, she wasn't in the same time zone or even in the same place. This was like running a relay marathon while frying pancakes. I got home, after picking up kids at kindergarten, when it was my turn, I made dinner, read a book in the sofa, watch the Sandman do his work, and then reopened the computer while attempting to relax slightly on the sofa, working until a few minutes past midnight. My last task of the day was a virtual meeting with the team before falling asleep like a pharaoh going to rest in a pyramid's tomb.

I've been in several of these kinds of groups and projects since. It is good energy and the work never slows down. The thing you notice pretty fast is how important that social call is – to have a meeting in the group and just talk about nothing. Letting all the non-verbal communication (to the extent that the digital environment allows you) be in focus. Sort out misunderstandings in spoken words and gestures. Talk about the kids or the problems of getting a decent cup of coffee in the US. More importantly, you learn to shut off your iPhone during the only quiet moment of the day. I can see you are not doing this as I glance at you from over my beer in 2019. You're rubbing that phone like you want a genie to appear. Give up! He is in my pint. No use trying to get him over to your side.

Of course, a social meeting is easier if you all meet in the same room. The conversation could be even more nuanced including body language, gestures, posture, tone and pitch of your voice, the clothes, smell and other aspects of face-to-face meetings. The impact of the natural language (see Figure 8.1) is pretty impressive and it is something you really notice when you are working in a completely digital team, totally cut off from the benefits of a physical meeting. Figure 8.2 provides practical tips for making a digital group work well.

Digital groups do it better

However strong the natural language really is, there are clearly advantages of working in a properly functioning digital team. First of all, I want to emphasise the advantage from a very planetary perspective: a digital meeting has a close to zero carbon footprint.

Think about a fly getting caught in a spider's web. The spider doesn't know that the fly is caught because it screams: 'I'm trapped, and I might be eaten!' The spider knows it has caught a fly because the web gets all wobbly and the lines get pulled. This is natural language. Body language – gestures, eye contact, posture – are examples of natural languages.

The tone and pitch of your voice could express stress or excitement without you thinking about it. Your clothing also sometimes expresses more than you have deliberately set out to express. The thing with natural language is that it does not lie. We therefore believe it more.

FIGURE 8.1 Non-verbal communication (natural language)

- Be aware of the differences between communicating from a distance and communicating in a room! Think about all the natural language that gets lost when connecting from a distance.
- A group discussion starts when everybody is present, both in the room and from a distance.
- Remember that comments written down and sent to group members by mail or in a digital forum tend to be perceived as harsher than if they are spoken by a person in the same room.
- Define formal and non-formal channels. One has to test, try and fail before succeeding in teamwork. It can be quite paralysing for creativity to work only in an official channel where everything is remembered and nothing is forgiven.
- Use the fact that nothing is forgotten. Then it is easy to dissect the full trail of the project and to analyse what went wrong and correct it.
- In a digital environment everybody can put in most work when it's optimal for them, thus creating a hive mindset of group members' best versions of themselves.

FIGURE 8.2 Easy tips to get a group to function well in a digital environment

The Mistra Center for Sustainable Markets (MISUM) is a multidisciplinary research environment at Stockholm School of Economics and has developed expertise on sustainability, since it's clearly hypocritical to talk about sustainability and then not walk the talk, or rather fly, and let down the environment. Digital is greener in many ways. We need to be careful about our world, and flying professors around to lecture on sustainability or food waste is a terrible idea. Let's connect from a distance like the TWC class does, uniting three lecture halls and let the students come together with technology, turning the world greener, instead of turning it into a green house. It is nice to live on a living planet, you know.

I honestly believe that projects with a group working properly on digital platforms perform better than the analogue groups, especially for mature professionals. Or more specifically, for individuals with a higher complexity in their normal day. One assistant professor told me after she finished a virtual meeting lasting an entire day with a dissertation group in London that she was so grateful that we upgraded one of the meeting rooms with webinar equipment. 'Now I can work for a couple of hours more, then go home to my family, instead of going to boring formal dinners in London. Oh, well, they are nice sometimes, as is travelling, and getting away from the family. But it's a good thing to have a choice.'

Yes, having the choice is always nice. Being able to work when it's the most convenient for you is grand. To study when it's optimal for you is fantastic. To listen to the course material when you are walking, exercising or frying pancakes, instead of reading it at a desk in a library. This doesn't mean that everything has to adapt to every wish of every single human being. But is it true that most people have a fighting chance in doing their best in an optimal environment?

'Some students do not look at the films and other material I put out for them to study before the classroom lecture', says Mattia Bianchi, a professor at SSE. 'Then they come to the first classroom lecture and they realise that they are in trouble if they have not.' Prof. Bianchi got the Best Teacher's Award at SSE in 2018. One student of his that I talked to confirmed that going to the lecture without watching the videos and looking at the other material is a mistake. Students hear about the class (it is popular) and the expectations (you cannot skip the pre-reading) which maximises the student experience since the learning starts even before the class begins.

We are still at the beginning of the digital journey at business schools, but I suspect that courses with a proper use of digital tools will on average score higher in student evaluations. When I was a teacher and programme director of the digital programmes at a marketing school – the first schools to get hit by the digital wave – the courses heavy on digital tools and material scored on average higher than the other formats. Why are the digital natives more satisfied with a course using suitable digital tools? In truth a mystery! Oh, all right. They might be used to digital tools and find it more convenient to work with them.

I must stress that these are my observations. The hard facts stem from course evaluations at one university and from my own observations made from my five years at The Royal Institute of Technology in Sweden and three years at Stockholm School of Economics (sources and material available on request). These are my own interpretations made from the dawn of the digital man. Also, please remember that digitalisation is for the most part an outside force. You can try and mould it into your own thing, but you can never choose to disregard it. Everything that can be digitised will be digitised. Sorry, but that's just the way it is!

Digital teachers do it better

Like my first fully distance project driven like a perpetual relay marathon of the walking dead inside a pyramid of frying pancakes that I mentioned a couple of pages back, a teacher with digital tools is now able to give constant feedback to a student. This is a crucial change from before. In a completely analogue environment, there is a high risk that the students work too long on the wrong track between meetings with the teacher. Throw-away work can be a difficult process if you have worked on it for a long time. With digital tools and the possibility of monitoring the work with shorter intervals, the risk of moving astray is lower. On top of that, the teacher can track the collaboration process and view the backlog, pointing out crucial events along the way (we could all use a Debbie in Texas at the end of each day telling us to shape up).

Another positive aspect of a digital work environment is justice. The group members' work commitment becomes more evident. It is harder to hide from work in a transparent digital environment where you can count responses and see turnaround time in minutes. Transparency is a word frequently used in the context of digitalisation. It is misused and overused to the brink of becoming a cliché, but transparency is a consequence of digitalisation, whether you like it or not.

A boss of mine a long time ago told me in the middle of a digitalisation project, when it became apparent that he was standing naked in a glass house that gradually lit up: 'You know I like transparency and we should be able to talk about *everything*. But just not about *this*.' He pointed mentally at a colossal failure that soon would be exposed. 'Had I known *this* [one thing], *this* [other thing] would not have happened.' If you are incompetent, transparency is not good; if you are a hard-working student, transparency – *this* – is a nice thing.

Transparency also punishes the teachers if they are not putting enough effort into the course, answering student questions or keeping up with relevant material. I had an easier task of sacking lazy incompetent teachers as a programme director of digital courses at the marketing school Berghs School of Communication than at my college working with the intensive course programme, held traditionally in a classroom. I could almost always find hard evidence to back up my accusation of not working hard enough with the students and with the course. Everything was logged.

Another beneficial factor of digitalisation is that a course undergoes a necessary redesign when it moves into a more digital environment, which almost always means updating literature and implementing better course structures. Or, as in the case of TWC, digitalisation made a completely new course format happen.

Official versus unofficial channels

If you are as old as I am, you remember the fall of the Communist regime, the tearing down of the Berlin Wall and the liberation of the Baltic States. If you are not, let me tell you this: it was very important in those days of upheaval and revolution to conquer the television stations to gain control of the official means of communication.

I know that this is confusing for a digital native, avoiding linear television with a natural grace. In a time when every brand is its own media house, it's very difficult to think about official and unofficial channels. But as a Danish waiter once said at my dinner party when asked if this was a smoke-free table: 'Smoke-free table? That obviously is something that you decide upon yourself.' (If you laugh at this, it means that you are old. Young people don't remember smoking indoors.)

What is an official and what is an unofficial channel? That is something that you decide. However, like smoke rolling over the tables of a restaurant in the old days, activities in one selected channel can prove unhealthy for you. I still remember discussions in 2008 when schools actually considered building online courses on Facebook and taking attendance through it. Your lecture on YouTube might be followed by a recommendation of a conspiracy theorist claiming lizards rule the world. You may not like that.

The university world has improved its official digital tools, supporting their own video services and webinar tools. The LMS (learning management systems) are getting more and more stable and no one would even consider using Twitter as a platform for their courses. Nevertheless, as the students are used to communication in private via a myriad of different channels, it could be counter-productive to force them into only collaborating on official university platforms. Give them some slack (a bit of pun intended) and leave them alone in their own channels. They are not going to share everything with the teachers. Also, give them unofficial digital channels on your university's digital campus. You might be able to check them but leave the comments behind for now. Wait until they have moved over to the official channel before giving remarks or comment.

Think about the fact that we are constantly under surveillance. Looking at our society today, it's clear that Orwell was an optimist. We are being watched and monitored beyond belief. Learn to look away and say to the students 'Whatever happens in this channel when you are working on your projects, we do not comment. It's when you move over to the official channel, I'm about to start my job of commenting on your work.'

Universities have an obligation to onboard their students, introducing official channels provided to them through their student credentials. It is not legal to force students to join Facebook in order to get course material or have group discussions. They will probably be on social media anyway, but it will cost you very little to give them a couple of extra, unofficial, channels as well to play around.

In the scope of the TWC course, we were very clear on labelling the channels that were provided to the students with official or unofficial. On the official channels the students knew that they were monitored by a teacher or tutor who could intervene, start a conversation or make direct comments in a shared document. The unofficial channels were off limits to comments. These channels are the student labs where drafts are written, and possible solutions are discussed. (A bit like the tiny apartment or the pub in the old days.)

Clear rules and consistency for each channel is of the essence! This logic is also found in app design where a good app normally is a tool to perform a specific task. If you would like to do something else, you simply go to another app. This mindset is pretty dominant with the digital natives: as an instructor or designer of a course you have to clearly define when the students should do what and why. It is especially important to specify exactly what the student should do to get a certain grade (read more about this in Howard Gardner's book *The App Generation*). Of course, this could prove difficult. We like to look at learning as having a wider scope than getting a grade or performing a task with excellence. So, while it's hard to define certain rules, it's smart and doable to define the rules of communication and presentation.

Collaboration in a digital environment is still a work in progress. It's like visiting a new continent. But instead of exploring it, we are inventing it day by day, inch by inch: truly a world in the making.

9

WHERE ARE WE GOING AND HOW DO WE GET THERE?

Sampo Sauri

Approach

An international course with students and faculty from three different universities in as many countries and students of over a dozen nationalities needs to pay special attention to how everyone can collaborate with ease. We had to think about the technology so the students could really work together over distance and stay connected regardless of physical location. Our goal was to find a solution where students could easily find the information they needed, meet and discuss with each other over long distances, and share documents and other material with each other. The students from different universities only met each other once before they were expected to start working as a tightly knit group. This meant that they had to quickly get to know each other in order to make it more natural to work together over a distance.

In 2017 we used Skype for Business for teleconferencing, Moodle for course assignments and discussion, but in 2018 we started using Microsoft Teams, which had all the requirements baked into one product, and the added bonus of having a mobile app for on-the-go communication. The need for Moodle as a separate platform dwindled, only to be left out in 2020 as the course increasingly relied on Teams.

Figure 9.1 shows the virtual classroom at Hanken with the partner universities connected through MS Teams.

User administration from a university standpoint

All students at the course need to be treated equally, thus we decided to have all students enrolled at one of the universities – in this case, Hanken. This was done for several reasons. Firstly, this way all students have access to the same tools, the

FIGURE 9.1 Virtual classroom at Hanken

same level of access to the course LMS and, if need be, access to the same journals and databases. Secondly, the students were enrolled at one university for accountability – this way, Hanken could effectively keep track of all student progress and prove that this was a course with intense international collaboration and have the ECTS credits shown in statistics. These metrics are very important from a funding standpoint.

This added bureaucracy did, however, not come without drawbacks. By being enrolled at Hanken, all students would of course receive a username and password to Hanken's online services, and to access the tools. In order to receive the credentials, all students had to verify their identities and sign Hanken's user agreement, basically vouching to not misuse their access to the tools and networks. All of this had to be done in person and monitored by Hanken staff, i.e. me. Upon signing the documents, the students would also receive the username and password required to log on to Hanken's services.

Each year, the course starts with a kick-off session where all student travel to the same place to meet up and get to know each other and go over the course assignments. Luckily, in 2017 and 2019 the kick-off was held at Hanken in Helsinki, so it was relatively easy to get everyone on board and sign the user agreements.

In 2018 it got tricky. The kick-off session was held at Stockholm School of Economics, and due to a jam-packed schedule, there was no time allocated for everyone to present their ID and sign the forms. As we made our way to the our corporate partner IKEA's store in Kungens Kurva outside Stockholm, I noticed a chance to strike: in a crowded IKEA bus – taking carless shoppers to IKEA and back for free – I started going from student to student, one by one, trying not to elbow the other passengers. IDs were checked, signatures were scribbled, and by the time we reached our destination I had collected nearly every student's user agreement and they, in turn, had received access to Hanken's tools and services. Success!

The following year the kick-off was held in Helsinki again, and I urged the organisers to schedule half an hour for the boring bureaucracy, which then went much more smoothly. As if that would not be enough, as per Hanken's policy all students need to change their passwords into something of their own choosing. This additional hurdle before they could start using the tools was a slight nuisance, but before long the students were all logged in and ready to go.

But as with everything, technology makes manual labour obsolete. From 2020, Hanken has evolved its processes so even international students could verify their identities and accept the terms online, using secure emails and forms. It's better for everyone this way, of course, but a part of me longs for the personal connection to the course, having to meet all the students brought with it.

Why Microsoft Teams?

When we first started the course in the spring of 2017, we used a combination of Skype for Business + Moodle, which meant that collaboration was more restricted. Of course, students used whatever tools they were used to, such as Google Drive, Dropbox and 'regular' Skype. The lectures were held over Skype for Business and compatible meeting rooms and auditoriums across the three universities, some students opting to attend using their own devices as well. This Skype for Business based set-up turned out to be somewhat cumbersome, and once Hanken had started using Teams in late 2017 the Tackling World Challenges course seemed the perfect fit for trying out Teams' many capabilities. It was a one-stop solution for video collaboration, sharing files and having a platform for discussion. Some students had used Slack previously, but most were new to this kind of a collaborative platform.

We set up Teams so there was one main team for the whole course (for all students, teachers and assistants), as well as one team per group of students. The main team was to share course information, for students to be able to ask questions, and we also had a channel set up for lectures. This meant that for each weekly or so lecture, students could come to campus and participate in class or meeting room through video conferencing equipment, or in case they could not attend in person they'd join online from anywhere using Teams on their computer or smart device.

The student teams only had student members, i.e. they could collaborate freely without the fear of a teacher eavesdropping on their discussion. This detail was important to us and had the aim of letting the students' creativity roam free and entice them to use Teams instead of fleeing to a more private platform. I believe we succeeded fairly well in this regard.

Additionally, there was one team for just the teachers and assistants – the so-called Back Office team, where teachers could discuss and collaborate, and quickly share information should something urgent arise. Figure 9.2 shows the set-up in teams.

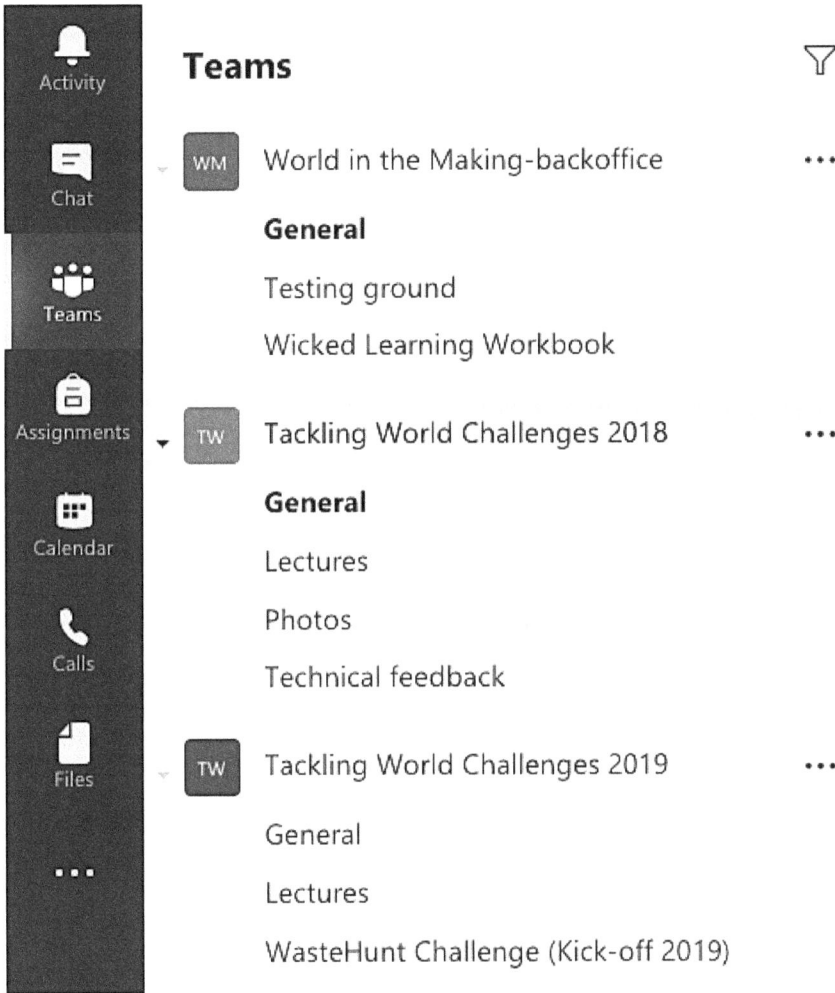

FIGURE 9.2 Example of MS Teams set-up

Since we started using Teams in 2018, the program itself has evolved and gained in features – in the course of 2020, there was only need for one Team, and the student groups got private channels to share and discuss in without fear of outside scrutiny mid-assignment.

Feedback on technology

Students seemed to find the technology used very efficient, but this also varied by group. Some students loved Teams and were surprised that it could integrate many features into one tool, while others preferred tools they were more accustomed to. The student groups were recommended to use Teams and the team created for the

Tell us what you thought of the technology we used

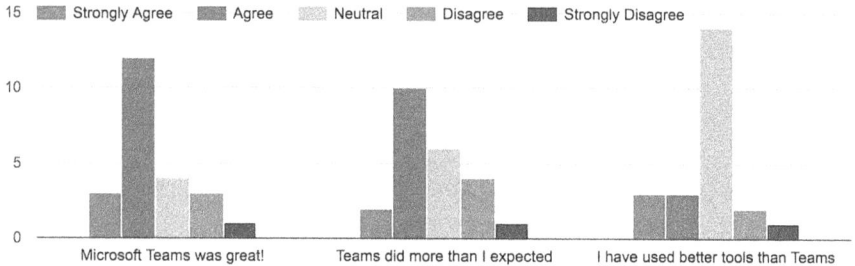

FIGURE 9.3 Student feedback re. technology used

group, but it was not something that we could enforce. Some teams chose to use other tools for collaboration instead, like Google Drive and Hangouts or WhatsApp and Skype.

In a post-course survey conducted in 2018 most students agreed or strongly agreed that Teams was great, and that it did more than expected. Overall, the students had also used other, comparable tools, but could not say if they were better or worse than Teams. Figure 9.3 shows the survey results. For the course's intents and purposes Teams performed quite well.

Conclusion

To sum up the experience from a technical standpoint: the only constant is change, especially when it comes to the tools used for collaboration. The technology has changed every year, become more versatile and inclusive and, arguably, easier to use. But as ever, our students have embraced technology to truly become an international unit, taking on challenges together without regard for distance or borders. They are an inspiration that's been a privilege to witness.

PART III

Method: Collaboration, tools, creativity, resilience, challenge, salt

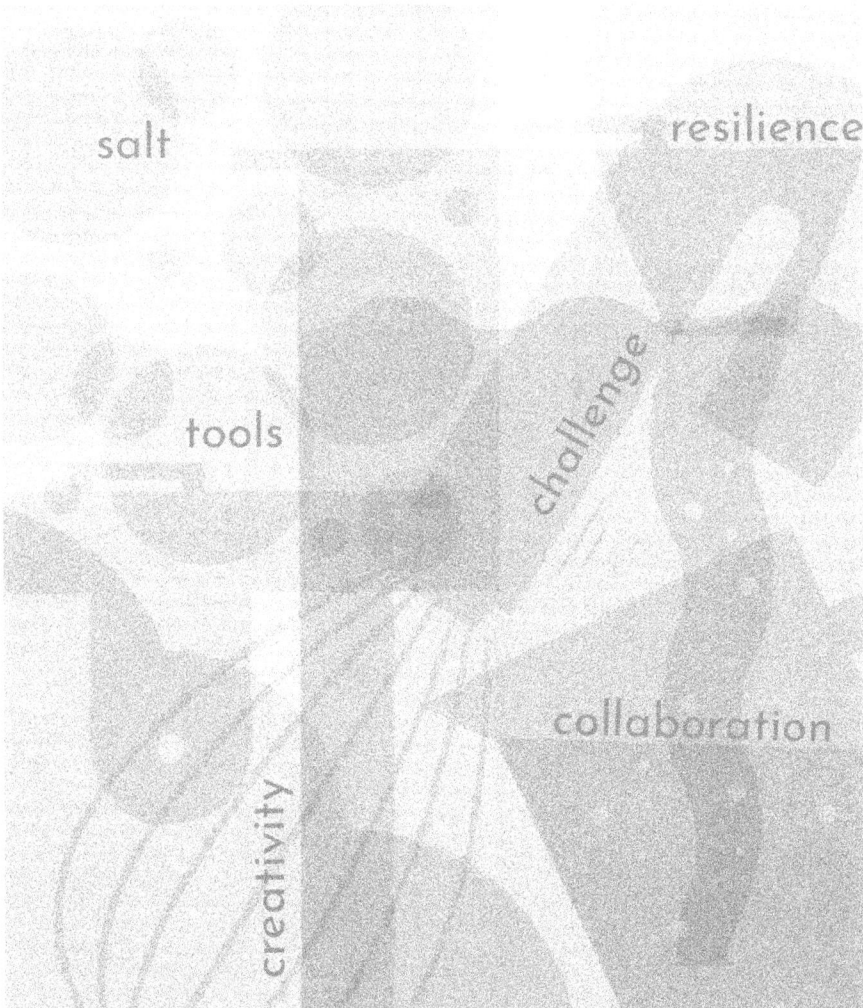

Drawing by Inês Costa

10

COLLABORATIVE PEDAGOGY AND OVERCOMING DIFFERENCES

Sofia John and Liisa Välikangas

Introduction

When we started this teaching collaboration, we potentially had the illusion that we knew what we were doing. We were providing our students with an international learning experience as well as the experience of working together with a corporate partner. However, not only were the students getting such an experience but so were the teaching staff. Running a course across different universities in different countries whilst also collaborating with a corporate partner posed a number of practical challenges. In this chapter, we will share some of these practical challenges.

Staying on top

'Hindsight is always twenty-twenty.'

(Billy Wilder)

It is not unusual when undertaking a project for the first time that one makes mistakes and learns as the project unfolds and once the project is complete. Be it buying a car for the first time, or travelling to a new destination, or starting a new relationship, few people have managed to get everything right the very first time. Often there are factors or consequences that we have been completely unaware of prior to undertaking the project. A new car-buyer, for instance, may have been unaware of the tax repercussions of purchasing a diesel car. Many things become much clearer in hindsight. When this course on Tackling World Challenges started, there were a number of unknowns. We had a general idea of the project the corporate partner wanted to give to our students: to cut food waste in a particular area of their business, and the students were to work together in teams to come up with solutions. The more specific details of the project would be ironed out as the project progressed. This was the first time we were collaborating with this corporate partner and how this project would translate into a student project was somewhat unclear in the beginning. Nevertheless, the timetable for the course had already been set before the course started. This meant that the faculty team had to set goals and deadlines for the project that fitted with the corporate partner as well as the course schedule. This demanded creativity and flexibility from all involved in the collaboration. Many elements of the course that were unclear at the time became unforgivingly obvious with hindsight.

As the project progressed, it became clear that the students had to come up with and present proposed solutions to the food waste issue, and then they had the chance to implement one of these. The corporate partner would decide which solution each student team would pursue. In hindsight, it was clear that we needed a session for the students to present their ideas to the corporate partner, but this was not obvious at the beginning. Thankfully, we had a spare session scheduled with no specific content, which then became the proposal presentation session. However, this issue and the decision to hold this presentation session came only a short time in advance. Furthermore, practical details, such as should all the students be present, or should the teams present individually in turn to the corporate partner, had not even been thought of, let alone discussed. Such details only entered our vision as the session approached and students asked about them. In this case, we decided that it made sense that the teams present individually to the corporate partner, and thereby they could limit the disclosure of their ideas to the other teams. Although decisions such as these pleased the students, these instances gave the course a messy feel because many details were decided and communicated to the students only a short time in advance. In other words, the course evolved as it progressed.

Furthermore, the students worked highly independently on their projects and with the corporate partner. Another challenge that the faculty team faced was staying in the loop with the students and the different stages of their projects. One team, for example, who were to implement their food waste solution in a Finnish subsidiary, faced difficulties because this subsidiary already had a protocol in place. This was unknown to the course faculty until the course was almost over. By this time, it was too late to think of and suggest an alternative for this team. Perhaps this team could have implemented their idea in another Finnish subsidiary? For some unknown reason, this option had not come up in the communications between the corporate partner and this particular student team. Perhaps if the local faculty had known about this situation, either this option would have presented itself, or an alternative project been suggested to the team. While better planning seems like an answer, it is also true that learning in complex situations is about coping with ambiguity. Here, a better result might have been achieved by being more aware of existing company protocols, yet a more important lesson may have been how to handle a situation that turns out to be difficult. When there are unexpected developments, there are also unexpected lessons to learn. Part of the course purpose is to learn to cope with ambiguity, complexity and uncertainty.

This in no way takes away the importance of being well prepared; quite the contrary.

Learning:

- Be aware that there are things that you don't know that you are unaware of at the outset.
- Ensure more regular feedback between students and staff.
- Learn to cope with unexpected developments and move forward.

Communication and coordination

'The devil is in the details.'

(Author unknown)

Often small things pose surprisingly big problems. The teaching collaboration occurred over two different time zones. Two of the universities and the corporate partner were in one time zone, and the third university was one hour ahead. It was surprising how such a small difference could cause such confusion. Although each of the faculty members had experience of working across border and time zones, we frequently had our calendars marked differently for scheduled staff meetings. Being explicit about what time zone a meeting is booked in is so obvious it almost does not warrant mentioning. However, such painfully obvious details are the ones that are often overlooked. On more than one occasion, someone from the staff team would show up either an hour early or an hour late to a scheduled meeting. Perhaps the fact that the difference was so small was what made it so confusing.

We had an unusually large teaching team with professors, teaching assistants, administrative staff, as well as technical support staff directly involved in running this course. Part of the reason for having such a large staff is the unusual nature of the course where students have to be registered in one university to get the course credit; there is a need for enabling and managing the virtual classroom and lectures; and the students also need have a local interface in their own university. The course could probably be taught with less faculty but an important purpose for the course is about cultivating inter-university collaboration. With such a large teaching team and a course that was ever evolving as it progressed, a crucial element of running the course was quick and timely communication amongst the staff members. Although we had more advanced technological applications available, the staff predominantly communicated via email. For the ever-evolving course, we faced challenges relating to decision-making. Should we wait for a consensus from everyone? Or was it enough to go ahead with a decision once a majority had replied? Who has the final word? There was no single university or person who had authority over the course.

Once a decision regarding the course was made, the next challenge was to decide who relays the information to the students and by what channel. Was it enough that one person posted in the information in Teams? We tried this but found that many of our students missed the information. In the end, we posted in Teams as well as sent emails to the students individually. One teaching team member from each university took responsibility to email the students of our respective universities.

Learning:

- Be explicit about details that are easily taken for granted, such as setting meeting times when there is a difference in time zone.
- Establish the best method of communication and a decision-making protocol at the start of the course.
- Have a definite leader for the course and its different activities (in particular communication).

Institutional differences

The EU has created a standardised credit system (European Credit Transfer and Accumulation System, or ECTS) in order to make degrees and courses amongst different EU institutions more comparable and transferable. However, each university has a different system for dividing and attributing courses and credits. Some universities divide smaller and bigger courses into four and eight ECTS credits respectively. Other universities divide them into five and ten ECTS credits respectively. Others may have a different weighting system. This was one area where we encountered difficulties. The course was supposed to be essentially the same for all the students irrespective of which university they were in, but for some students this course was worth eight ECTS credits, and for others it was worth six ECTS credits. This raised questions about fairness amongst the students. How could some get eight ECTS credits with the same work as others who only got six ECTS credits? This issue was addressed with the second major institutional challenge we faced: differing examination requirements.

Not only do universities differ in the way they set ECTS credits for courses, but they also differ in other rules and standards pertaining to running a course. We found that our three schools had different examination requirements. One university required that the final grade have a maximum of 50% group work and a minimum of 50% individual work. The other two universities had no such rules, but in order for the course to be equal cross-university, the entire course had to be designed so that the teamwork constituted 50% of the grade, and an individual essay or exam constituted the remaining 50%. Each university was individually responsible for the individual assessment. This also made it possible to address the issue of the different credit-weighting amongst the universities. Those with a higher ECTS credit-weighting could demand a more substantial individual assignment, and those universities with a lower ECTS credit-weighting could have a lighter individual assignment. Nevertheless, this was a weakness of the course that tended to sow discontent amongst the students. Subsequently, we have made efforts to clearly state examination requirements up front, and openly acknowledged institutional differences. Our aim is to develop a single assessment method and ensure that the course credits and examination requirements are the same for all students. Meanwhile, the assessment requirements are tailored to each university's requirements such as an individual exam.

Learning:

- Be aware of institutional differences.
- Institutional differences can even benefit the collaboration in that the course builds on the varying requirements of both individual and team performance.
- Communicate the assessments and examination particularities up front so that the students understand the reason for the differences (and can take them into consideration themselves in their own performance).

Conclusion

It is said that well-planned is half-done. But a course like this also requires learning-by-doing. This chapter shares some of those learnings from running the course. Our learnings are akin to the so-called lean product development where the first step is experimenting on a Minimum Viable Product (or MVP). An MVP is a launch that is not perfect, far from it, but hopefully a version that can survive and get better as more experience is gained. We now focus on improving the live course based on our cumulative experience. This is true to our Wicked Learning approach: be courageous and learn fast.

11

A TOOLKIT FOR TACKLING WORLD CHALLENGES

Approaches and methodologies for teaching sustainability

Marijane Luistro-Jonsson and Anna Nyberg

Introduction

Teaching how to tackle sustainability and the challenges surrounding it, particularly in a business school, is a challenge in itself for various reasons. Amongst others, the multidisciplinary nature of sustainability requires teachers to educate students beyond their comfort zones. There is a need to build literacy and expertise in other academic disciplines, shift perspectives, guide critical thinking and develop 'change-making' skills and 'championing' capabilities to enable students to understand and act on the challenges. These broad-oriented and socially relevant outcomes are to be ambitiously achieved in educational institutions that commonly espouse traditional homo-economicus paradigms. Moreover, there is no clear-cut template nor will there ever be a recipe on how to teach future leaders and decision makers to save the world. The array of management tools and economic models are often oriented towards dealing with tame problems, not the wicked problems covered in sustainability courses. Thus, designing the content of sustainability courses has generally been a perennial challenge in itself, requiring new pedagogical approaches and learning tools.

Unlike other sustainability courses, this elective master-level course has practical, internationalisation and digitalisation components in addition to its sustainability content. The practical real-world connection allows the students to develop an understanding and apply their knowledge as they work with one of the UN Sustainability Development Goals alongside an industry partner. The international setting gives the exposure of working in globally dispersed teams, and helps develop skills in collaborating with people from different cultures. The digital technology enables the virtual groups to use new technology in bridging geographical distance and exploring new learning techniques. Given this rich learning ground for experiential learning, we teachers took the role of being the facilitators of learning rather than experts of knowledge. Instead of imparting a proven approach, what we thought fitting to the situation was to provide the students with a toolkit that helped them to magnify the nature and intricacies of the problems and to equip them with various lenses that aid in prescribing their own diagnosis and measures to problems with no clear-cut solutions. The toolkit is also a means for them to connect and apply the knowledge and methods they have learned from their other courses and to create a bridge to the working world.

The remaining sections of this chapter describe the contents of the toolkit used in the course. We begin with presenting the set of components as a whole, then discuss how each individual component helps the students in crafting their problem-solving skills for sustainable challenges. Finally, we conclude with the insights we have gained.

A toolkit for Tackling World Challenges

Figure 11.1 gives an overview of the different components comprising the toolkit for Tackling World Challenges. In general, we included the theoretical tools providing perspectives of the problem, skills-development tools working in teams and with industry partners, the various direct and indirect methods in designing, implementing and evaluating solutions, and the skills development and digital technology tools. We are aware that other components can be suitably included but the current discussion depicts what we have only so far included in the course.

Understanding the challenge

The students begin their journey in the course by being equipped with tools helping them to understand the challenge, particularly the *nature of the problem* and the *current trends and issues* surrounding it. As mentioned, the challenges involved in this course do not share the same characteristics as 'tamed' problems (i.e., it is clear if the problem has been solved or not) traditionally covered in other courses. These societal and global challenges require solutions going beyond management matrices, statistical analyses and economics models supporting decisions on how Company A can increase their return on investments or which strategy is most feasible for Company B to take. Thus, we find it important to familiarise the students with the concepts of 'wicked problems' (Rittel & Weber, 1973) and 'grand challenges' (Ferraro, Etzion & Gehman, 2015) to help them to re-orient their level of analyses and expectations. As these concepts elaborate on the complexities and ambiguities behind these types of problems, it gives the students a realistic view that they are not out to solve the world's problems but are contributing to resolving them. By doing so, we also move beyond the sustainability rhetoric, digging into the nature of the problem and its complexities, and not just presenting the challenges as part of a normative international agenda.

In addition to understanding the nature of the challenge, we also cover the research forefronts of the challenge involved by inviting industry experts and

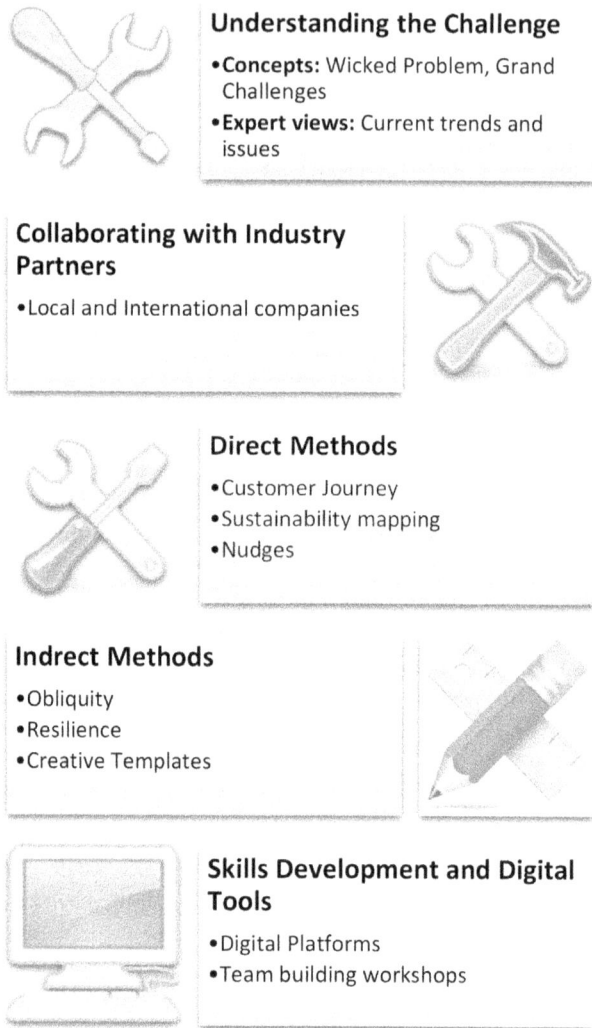

Understanding the Challenge
- **Concepts:** Wicked Problem, Grand Challenges
- **Expert views:** Current trends and issues

Collaborating with Industry Partners
- Local and International companies

Direct Methods
- Customer Journey
- Sustainability mapping
- Nudges

Indrect Methods
- Obliquity
- Resilience
- Creative Templates

Skills Development and Digital Tools
- Digital Platforms
- Team building workshops

FIGURE 11.1 Toolkit for Tackling World Challenges

leading researchers to get an overview of current trends and issues. Invited guest lecturers from leading food-waste think tanks like the International Food Waste Coalition (IFWC) and EAT Foundation have enlightened the students of the severity, impact and layers of wickedness of the challenges. For instance, the EAT Lancet report (Willett et al., 2019) related the food-waste challenge to dietary issues and planetary boundaries, showing where food-waste reductions fit in shaping sustainable food systems that are within safe environmental limits. Gaining perspectives on how the challenge connects to the broader picture is essential for students to come up with more informed solutions.

We focused on the same challenge of reducing food waste for the past three years. Sticking to the same theme allowed us to follow the developments in the field, build on cumulated experience and knowledge and gave enough time to tackle the different facets of the issue – from food waste in the farm, the supply chain and on the plate of consumers. Having different themes could have equally worked, and the decision depends on how it fits with the next tool, which is the industry partner.

Collaborating with industry partners

Giving a context to the challenge is another essential part in setting the scene of this course. To find companies which are willing to let the students gain a first-hand experience to work with such messy issues has been valuable for the course. We have experimented working with different partners through the years, which enabled us to get an insight how different types of organisations have been working with the food-waste problem. The students worked in national teams with school canteen food suppliers, small local restaurants, food delivery start-ups and pre-schools, as well as in international groups with leading multinational retailers and manufacturers who are directly and indirectly involved in the food industry. Having industry partners who have international operations has been an advantage as their standardised facilities and processes (e.g. restaurant chains) allowed the students from the different countries to implement similar solutions, gaining insights on the cultural differences as they compare and contrast the experiences and results.

What we commonly find to be important on the side of the partners is their openness to share their current strategies, time and resources to supervise the students, and respect for the ideas and capabilities of the students. On our part, advance coordination and engaging partners' interest is important. Approaching potential partners and finding the right one is a journey in itself, marked with last-minute changes and finding of new partners.

The industry partners play an important role in shaping the course as they provide the context on what the students can work with, defining the scope of what they can and cannot do about the challenge. In some cases, the students were given full access to information and resources necessary for them to explore their ideas and were provided with internal supervision from the companies. In other cases, they did not have access to some information and were denied the opportunity of pursuing some ideas. Regardless of the project objectives and space allotted by the industry partners to the students, they played the role of learning facilitators equally as they provided the professional settings to actualise the ideas and capabilities of the students.

Direct approaches

A series of lectures, spearheaded by the different schools, discussed several indirect and direct approaches and methods that can aid the students in designing their projects before embarking on them. Direct approaches entailed the process from ideation to evaluation of solutions, and included explorative methods such as customer journey and materiality mapping, as well as experiment methods testing interventions like nudging. These tools aim at directly finding the causes of the problem and creating solutions to address them.

The starting point of the direct methods is to explore where and why the problem exists, whether the context is food waste in the supply chain, restaurant consumers or households. A customer or product journey map is a tool that visualises the different interactions and decision points of the actors involved with food waste – and that includes tracing their actions, motivations, questions, emotions and barriers (Richardson, 2010). The students were able to do this through, for example, interviewing and observing restaurant and canteen customers, as well experiencing being in their shoes themselves. Another way in aiding them to understand the problem from the perspective of the company is to conduct a materiality map, where one identifies how the sustainability issues relate to the key stakeholders and the strategies of the company (NYU Stern, 2017). These mapping tools, in addition to the literature reviews, interviews and surveys conducted by the students, became the analytical foundation from which they developed and tested their unique solutions.

Reducing food waste inevitably entails behavioural change, whether at the individual, organisational or governance level. As such, a number of the projects dealt with testing various nudges or interventions that changed the choice architecture and influenced people's behaviour in predictable ways without changing economic incentives or forbidding any options (Sunstein, 2014). The experiments ranged from engaging and informing children through games and activities, to having subtle information in trays, posters, serving counters, leaflets and tables, as well as to using social media to influence behaviour. The interventions were pilot attempts allowing the students to test their experimentation skills. Even if ideal measures were often not available to help them evaluate the full impacts of their interventions due to time constraints, it provided a platform for students to design and implement their novel ideas, familiarising them with the process. Although the measured effects were not sufficient to give generalised conclusions, they nevertheless gave indications of the direction of the results. The whole process of enabling the students to have a grounded analysis and test their ideas in the real world empowered them to tackle directly the challenges in their own ways.

Indirect approaches

The concepts of obliquity, creativity and resilience acquainted the students how to approach the problem alternatively or to look in another direction. During the initial lectures, the notion of obliquity, which argues that complex objectives goals can be best achieved when pursued indirectly (Kay, 2011), was highlighted. Moreover, creativity templates for problem-solving, which emphasised importance of associations, divergent and convergent thinking, and attention disengagement gave tips on how to come up with more unusual and original solutions. Thus, it was enriching to expose the students to other cases before embarking on their projects with the industry partners. In one of the schools, the curriculum included local field trips, where the students had to make reflections on how other actors (e. g. cruise ships, municipalities) handled food waste, giving them exposure to related problems and solutions from which they can associate their ideas.

As a result, we have seen some groups who have chosen to investigate the problems from other perspectives, not from the operations and confines of the firm or industry partner. For instance, one group analysed the contracts of the school and the municipalities instead of working directly with the food caterer and management service provider of the canteen. Another approach is doing detective work into the resiliency of the organisation, and getting a retrospective idea on solutions on how to handle crises related to the challenge. For example, groups analysed how food-waste incidents led to higher reliability due to collective mindfulness (Weick, Sutcliffe & Obstfeld, 1999).

The fine line between the direct and indirect tools applied in the course can, however, be quite blurry. For instance, the direct experimentation tools can also work indirectly such as in the case of using mindfulness amongst the students to reduce their food waste. Alternatively, indirect methods such as analysing the contracts between municipalities and schools actually hit on the core source of the problem. Regardless of how the tools were used, directly or indirectly, offering a wide variation and different approaches allows students to create innovative solutions without them replicating each other.

Skills development and digital technology

The course also needed tools to address internationalisation challenges. The diversity amongst the students has been one of the strengths of the course, yet it also poses risks for conflicts and that richness is not being tapped into. There were workshops given to aid the students in how to work in international groups, as well as how to effectively communicate virtually. The workshops aimed to develop not only group cohesion but also communicating skills in a virtual world.

The digital tools included in the toolkit pertain to the digital platforms such as Teams which enabled the three schools located in different geographical locations and having different time zones to conduct the class together. The platform also allowed the students to work with each other and connect with the companies, which suited everyone's schedule. As pointed out in the other sections of this book, adjusting to this new way of teaching and learning has its ease and difficulties for everyone involved, but has nevertheless made it possible for the course to emerge and evolve.

Conclusion

The pedagogical approach of providing an expanded problem-solving toolkit to tackle world challenges gave the needed flexibility and resourcefulness for the students to come up with innovative and promising solutions. They did not seem to have any problems in deciding which tools to use, and in the end a variation of interesting solutions emerged. This approach matches the complexity and wickedness of the problems involved as it offers different problem-solving techniques across disciplinary boundaries, and develops the skills and capabilities of the students to work in geographically dispersed teams, with projects in different locations as well. Taken together, the projects collectively show that no single approach works and there is a need to continuously improve and update the contents of the toolkit. There is also a need to sharpen the theoretical lenses, refresh methods and add new ones according to the needs of the new partners.

There are inevitable challenges to providing a toolkit approach. For one, students think that the course is unstructured, they might find the content less rigorous compared to other courses, and they might not have the tenacity to deal with the ambiguities involved. Nonetheless, we think that taking this risk is worthwhile because as the students enter their professional lives they will face similar wicked problems to solve and they will similarly work with the toolkit of knowledge they bring with them.

References

Ferraro, F., Etzion, D. & Gehman, J. (2015). Tackling grand challenges pragmatically: Robust action revisited. *Organization Studies*, 36(3), 363–390.

Kay, J. (2011). *Obliquity: Why our goals are best achieved indirectly*. London: Profile Books.

NYU Stern Center for Sustainable Business. (2017). *Materiality and sustainability strategy for business*.

Richardson, A. (2010). Using customer journey maps to improve customer experience. *Harvard Business Review*, 15(1), 2–5.

Rittel, H.W. & Webber, M.M. (1973). Dilemmas in a general theory of planning. *Policy Sciences*, 4(2), 155–169.

Sunstein, C.R. (2014). Nudging: A very short guide. *Journal of Consumer Policy*, 37(4), 583–588.

Weick, K.E., Sutcliffe, K.M. & Obstfeld, D. (1999). Organizing for high reliability: Processes of collective mindfulness. In R.S. Sutton & B.M. Staw (Eds.), *Research in Organizational Behavior*, Vol. 21 (pp. 81–123). Stanford: Jai Press.

Willett, W., Rockström, J., Loken, B., Springmann, M., Lang, T., Vermeulen, S. … & Jonell, M. (2019). Food in the Anthropocene: The EAT–Lancet Commission on healthy diets from sustainable food systems. *The Lancet*, 393(10170), 447–492.

12

WICKED PEDAGOGY AS CREATIVE BRICOLAGE

Michael Gibbert, Monika Maślikowska and David Mazursky

Introduction

> And so work has changed its character. More often than not, it is a one-off
> act: a ploy of a bricoleur, a trickster, aimed at what is at hand and inspired and
> constrained by what is at hand, more shaped than shaping, more the outcome
> of chasing a chance than the product of planning and design. (Bauman, 2000,
> p. 139)

If you are interested in applying the above pedagogy, at some stage the question of
resources will come up. In this chapter, true to the 'wicked' spirit of the book, we
take a step back from the usual idea that we need abundant or at least adequate
resources to implement new ideas and instead argue that it is actually scarce
resources that drive innovation – in teaching sustainability and elsewhere.

To begin with the 'elsewhere', we are drawing on our joint work in consumer
creativity and problem solving more generally. We will argue that scarce resources
(even in the form of dysfunctional products) are actually instrumental in bringing
forth breakthrough innovation via 'bricolage', i.e. unusual combinations of previously separate objects, products, and concepts more generally. This research (on
which we have published several articles, special issues, books, and so forth: see
Hoegl, Gibbert & Mazursky, 2008; Gibbert & Mazursky, 2009; Gibbert &
Mazursky, 2007; Gibbert & Scranton, 2009; Estes, Guest, Gibbert & Mazursky,
2012; Gibbert, Hampton, Estes & Mazursky, 2012; Gibbert, Hoegl & Välikangas,
2014; Välikangas & Gibbert, 2015) is, we believe, particularly relevant in the context of sustainability, which is inherently about bricolage, i.e. saving resources,
making do with what is at hand, not wasting anything. As we will show in this
chapter, these sustainability mantras, far from being an end in themselves can be a
means to an end, and the end is innovation.

Teaching sustainability as bricolage

There are some striking parallels between pedagogical innovation and product innovation. Almost exclusively, research on product innovation including disruptive innovation, user innovation and consumer creativity has focused on understanding how users or other firms along the vertical/horizontal chain create new ideas based on products that work, i.e. function properly (e.g. Markides, 2006; Schreier & Pruegl, 2008; Herstatt & von Hippel, 1992; Kristensson, Gustafsson & Archer, 2004; Dahl & Moreau, 2005). However, in many usage situations, users are faced with products whose functionalities are disrupted to varying degrees (e.g. batteries of portable computers do not recharge any more, pre-paid mobile phones run out of credit, etc.).

Similarly, the idea for this course was borne out of a product (a.k.a. pedagogy) that doesn't work anymore. If we want to prepare our students to work in the new working conditions (Maślikowska & Gibbert, 2019) of virtual, interdisciplinary and international teams working in dynamic participation rather than the classic bounded membership (Mortensen & Haas, 2018), traditional real-space classroom-based models fail to deliver. In fact, they do not even approximate the functionalities and skills we need to impart to our students (teamwork, cross-disciplinary and international collaboration – see Chapter 7 in this book). In these times of education facing the global phenomenon of internationalisation (Hardy & Tolhurst, 2014) and business school curricula introducing global sustainability issues, academia needs to address the associated complexity (Aragon-Correa, Marcus, Rivera & Kenworthy, 2017), including an increased diversity and multidisciplinarity (Hardy & Tolhurst, 2014). Moreover, research on teaching sustainability in business education shows that in order to face the challenges of sustainability we need to educate students to become integrated catalysts (Akrivou & Bradbury-Huang, 2015), which can be achieved through different means, including the application of *phronesis* (Antonacopoulou & Bento, 2018). Such an approach to education engages not only academics but also business practitioners and students in the co-creation of knowledge (Antonacopoulou, 2010) to become 'Leaders-as-Learners' (Antonacopoulou & Bento, 2018) allowing the students to innovate in an ad hoc manner and learn on the go, while working in the context of uncertainty.

Let's turn to product innovation, for illustrative purposes. Here, functional impairment forces users to cope creatively with the functionally disrupted product by finding a new, ad hoc category for it and innovating its use for the new purpose (e.g. Upper, 1974; Hirschman, 1980; Barsalou, 1982, 1983), a process both popular parlance as well as academia refers to as *bricolage* (Lévi-Strauss, 1966; Weick, 1993; Duymedjian & Rueling, 2010). Even though bricolage constitutes an integral part of the lives of everyone, almost on a daily basis, research on teaching sustainability is all but silent on the factors influencing this process. In this chapter, we review theories from anthropology and cognitive psychology to shed more light on the factors impacting our information-processing strategies during bricolage. The importance of bricolage in new product innovation and in creative tasks more generally has been established first in Claude Lévi-Strauss's 1962 book *La Pensée Sauvage* (published in English as *The Savage Mind* in 1966), and has seen growing impact in the social sciences, including anthropology, cognitive science, information technology, entrepreneurship, innovation research and organisational theory (Duymedjian & Rueling, 2010, for a review and extension), and we believe it is essential to understand (and extend, i.e. adapt to your own purposes) the present pedagogy.

Some conceptual background on disruptive/disrupted innovations and bricolage

Disruptive innovations vs. disrupted innovations

There is a twist. Disruptive innovation is a core concept for innovation management based on thinking by Austrian economist Joseph Schumpeter from the 1940s regarding 'creative destruction', and suggests, in a nutshell, that 'real' innovations tend to destroy or at least disrupt the status quo (e.g. Markides, 2006). The twist is that disruptive innovation continues to be typically viewed in the context of existing products that function properly. Despite this, dealing with products that are characterised by functional disruption has, for better or worse, become almost an integral part of our daily survival.

Researchers have categorised innovation in dichotomies, such as radical versus incremental (Albernathy, 1978), competence-enhancing versus competence-destroying (Tushman & Anderson, 1986), architectural versus modular (Henderson & Clark, 1990), and sustaining vs. disruptive (Christensen, 1997; Christensen & Raynor, 2003; Christensen, Anthony & Roth, 2004), with the last being particularly relevant for our purposes here. In comparison to sustaining innovations, disruptive innovations effectively establish a new market by launching a new kind of product or service (or modifying an existing product), and which (somewhat counter-intuitively) are less desirable to customers along one or two key dimensions. Sony's early transistor radios, for instance, sacrificed some of the radio's sound quality but offered a new and different package of attributes: small size, light weight and portability, thereby creating a new market for a modification of an existing product (Bower & Christensen, 1995).

While existing work in marketing and innovation management does not distinguish directly between slight and severe functional disruption, empirical evidence in cognitive psychology suggests that functions differ with regard to their centrality. Specifically, a distinction is made between minor functions and core functions (Sloman, Love & Ahn, 1998). A function is core to an object to the extent that people (1) claim that they would be surprised to encounter the object without the function, (2) claim that imagining the object without the function is difficult, and (3) rate the object without the function as atypical of the object category (Sloman, Love & Ahn, 1998, p. 191). The distinction between minor and core function underlies further categorisation of disruptive innovation into typologies, depending on which of the two functions is manipulated.

Christensen (1997) points out two distinct types of disruptive innovation: low-end and new-market. In the case of a product or service that is 'too good' (offering more than customers want or need) and, as a result, priced higher than the market is willing to sustain, a 'low-end disruptive innovation' removes (unnecessary or unwanted) functionality so that the price can be reduced to a level that the market accepts. The massive discount retail chain Walmart exemplifies low-end disruptive innovation – they opened their doors with 'no-frills shopping' for basic goods and the lowest prices in town – catering to customers' preference for 'less good' and also less expensive products. Conversely, new-market disruptive innovations remove an undesirable minor function which had limited the potential uses of a product. This appeals to customers and results in more sales, thereby expanding market share. There are many examples of new-market disruptive innovations, including mobile phones (which eliminate the need for a landline), personal computers (and then laptops) as a desktop solution for computing (reducing a customer's dependence on mainframe computers). These examples both illustrate moving into a new market (from the original one), and because customers were given access to something that previously required significant investment or expertise, the new market also experienced new growth (Christensen, Anthony & Roth, 2004).

Let's summarise. It appears that existing work has mainly focused on (a) downgrading a core or minor function (low-end encroachment / low-end disruptive innovation, as in the case of the progressive reduction in hard-disk diameter (while having less storage capacity, the smaller disks were not only cheaper and more rugged but also used considerably less power, see Bower & Christensen, 1995, pp. 45–46)); or (b) the extinction of an *un*-desirable minor function (new-market disruptive innovation, as in the case of cable in the landline phone to come out with the mobile phone). Overall, then, nothing much is known about the complete extinction either of a core function (leading to severe functional disruption of the original product) or of a desirable minor function (leading to slight functional disruption). For instance, what would happen if the voice function disappeared from a telephone? Or the possibility to print from a printer was eliminated? What if students on a course would not sit together in the same classroom at the same time when being taught?

Turning to marketing theories helps tackle these questions: traditional theories such as conjoint analysis or discrete choice analysis typically look at products as bundles of functions, attributes and features (Olson & Thomas, 1983). Similarly, traditional pedagogy in universities is based on a bundle of features which are rarely challenged such as the idea of the professor (in the UK still referred to as 'reader') providing a 'lecture' which the students are supposed to understand because they have (in theory) prepared the chapter/lecture at home in self-study. The Flipped Classroom is perhaps the single most disruptive innovation hitting pedagogy since the first universities were founded in Al-Karaouine (859), Bologna (1088), and Oxford (1096). It is based on Eric Mazur's book from 1997 (*Peer Instruction: A User's Manual*). The title 'Flipped' classroom says it all: it disaggregates the bundle of lecturing in class and studying at home in that it inverts the tasks that are done at home and the tasks that are done in class. Traditionally, you ask students to read the relevant book chapter or other readings in preparation for class. The professor (or 'reader') then uses class time to regurgitate what the students should already know, recently prevailing upon hundreds of PowerPoint slides over the course of a semester. As my co-author Patricia Stokes from Columbia University and I write in our new book (Stokes & Gibbert, 2020), professors prioritise only certain parts of the voluminous and chunky text, which in effect reduces students' motivation to prepare for class since it is counter-productive—students realise that only some key parts will be presented so why waste time studying the entire chapter assigned? (see my ruminations in Stokes & Gibbert, 2020, pp. 37–38). An innovative alternative is to flip the classroom, turning class sessions into interactive time where concepts are applied to real problems from cases or local examples together (with the professor and in groups). Students prepare for class by learning the required concepts via pre-recorded lessons (different from MOOCs because they are the professor's videos on the subject only available to the students of that particular class). This method engages students in becoming responsible for their own learning through hands-on exercises and real-time practice. Students cannot get away with hiding behind their devices or catching up on other work, possibly unrelated to the subject at hand. Instead, students are right in the middle of their own classroom experience (dethroning the professor), which similarly to the previously mentioned *phronetic* approach, allows students to become 'Leaders-as-Learners' (Antonacopoulou & Bento, 2018) and the integrated catalysts (Akrivou & Bradbury-Huang, 2015). So, time at home is dedicated to good old-fashioned studying (where the students themselves need to prioritise the info they take in) and in class, there is a reality check that the (most relevant) material has actually sunk in. The course we are offering integrates these approaches, addressing the emerging malfunctionality of the current educational system, deleting some of its core features (time-space contingencies in classroom-based teaching in one university), and allowing students to interactively take part in adapting to the new context and reality (Maślikowska & Gibbert, 2019), including virtual collaboration and dynamic team participation (Mortensen & Haas, 2018), diversity and multidisciplinarity (Hardy & Tolhurst, 2014), sustainability (Aragon-Correa, Marcus, Rivera & Kenworthy, 2017) and wicked problems (Rittel & Webber, 1973; Ferraro, Etzion & Gehman, 2015).

The influence of bricolage on functionally disrupted products

Coming back to product innovation: making something out of nothing, i.e. making do with whatever functionally disrupted product is at hand and creating new forms and order from tools, materials and resources that are readily available (as opposed to procuring purpose-built tools from elsewhere) has been defined by anthropologists as 'bricolage' (Lévi-Strauss, 1966). Arguments regarding the influence of bricolage on the evaluation of the developed solutions are based on the existence of a so-called 'path of least resistance'. The underlying theory suggests that when facing a problem, the default approach of people is to implement the first solution that comes to mind and seems to fit based on prior experience (Ward, 1994; Ward, Dodds, Saunders & Sifonis, 2000). In Lévi-Strauss's original work, this approach has been linked to the thinking process of what he calls the *ingénieur* (engineer), who would have the knowledge about tools specifically adapted for his or her project, as well as the financial means to procure them. The *bricoleur*, by contrast, draws on what Lévi-Strauss calls his stock, which may often seem odd and heterogeneous, since it typically is not specifically adapted to the project. Nevertheless, the *bricoleur* has kept resources, tools and materials, even after they served their purpose, as they might come in useful at some point (Duymedjian and Rueling, 2010, p. 137). Thus, while the *ingénieur* strolls along the path of least resistance by procuring tools and resources specifically adapted to the project, the *bricoleur* leaves the path of least resistance, charting unknown territory by using whatever is at hand, adapting it to the current project in creative ways.

The pedagogy we espouse here is meant as a gentle nudge to push students off the path of least resistance when it comes to teaching sustainability. To stray from this path of least resistance demands not only a higher level of uncertainty in the outcome but also more cognitive effort (Ward, 1994). Only when established approaches are not available, e.g. because they are functionally disrupted, are people more likely to seek truly novel solutions off that path of least resistance. This is illustrated in prior work on use innovation (Hirschman, 1980), which suggests that consumers make use of functionally disrupted products in new ways. The general approach is to seek solutions within the boundaries of the current object set, rather than in the marketplace (Leonard-Barton, 1992; Ward, 1994). Specifically, consumers may use (or shall we say ab-use) a functionally disrupted product in a way that was *not* originally intended, thereby potentially creating a new 'ad hoc' category for that product (Barsalou, 1982, 1983). For example, using tin cans that previously held vegetables to store nails in the workshop, rather than buying containers specifically designed for that purpose (Ridgway & Price, 1994, p. 76). When there is a problem and no pre-existing solution is readily available, consumers use 'what is at hand' to construct a new solution (Finke, Ward & Smith, 1992; Hoegl, Gibbert & Mazursky, 2008; Dahl & Moreau, 2005). For instance, consumers may use the display light of their mobile phone to find a keyhole when caught unawares on a dark staircase (i.e. using the phone as a substitute for a flashlight; Christensen, Anthony, Berstell, Nitterhouse, 2007). To summarise, bricolage refers to situations in which consumers seemingly 'misuse' a functionally disrupted product, i.e. use it for a purpose for which it was not originally made, thereby changing the category of the old product, and turning the old product into a successful 'new' product (Hirschman, 1980, pp. 292–294; Ridgway & Price, 1994, pp. 76–80). In our case, limited resources (financial, above all) make it impossible for students of three universities to sit in the same physical classroom at the same time. What do we call a university's classroom that is no longer a classroom in a university?

Theoretically, bricolage should be positively related to the evaluation of the outcome in the case of products characterised by severe functional disruption, but not necessarily in the case of slight functional disruption. Recall that bricolage refers to situations where the product is to be used in new ways, i.e. a new, 'ad hoc' category is to be formed. Work in cognitive psychology has shown that due to the higher amount of cognitive effort it requires, such re-categorisation is not automatic, and individuals actually resist doing so, as illustrated by the reluctance to leave the path of least resistance discussed above. Since slight functional disruption is defined as the extinction of one of a product's minor functional features, the absence of such a function does not affect the product's original category coherence. A bicycle without a bell is still a bicycle; a GPS navigator without the option of choosing whether a male or female voice gives the directions is still a navigator, for instance. However, in the case of a core function, people claim that they would be surprised to encounter the object without the function, claim that imagining the object without the function is difficult, and rate the object without the function as atypical of the object category (Sloman, Love & Ahn, 1998, p. 191). Thus, few people would say that a bicycle without wheels is still to be considered a viable bicycle, and the same would be said of a GPS navigator without the navigator function. By definition, the disruption of a core function therefore severely disrupts a product's very essence, nudging individuals off the path of least resistance, and forcing them to creatively consider what they have (left) 'at hand' in terms of its appropriateness for alternative, novel ends. In the innovation and new product development literature, work on so-called 'creativity templates' suggests that eliminating such a core (rather than a minor) function from a product's configuration may even be a means for creating new, successful products (Goldenberg, Lehmann & Mazursky, 2001; Moreau, Markman & Lehmann, 2001).

Note again the conceptual difference between bricolage and traditional user-driven innovation as well as consumer creativity research: while bricolage explicitly refers to functionally disrupted products, user-driven innovation operates within the realm of products whose original core function is used for a different purpose, precisely because it is still 'working'. Consider Viagra and Botox, two products that are often cited as prime examples of user innovation. The very core function underlying both products is preserved but used for different means. If that function were no longer present, Viagra could not be used to treat ED, and peoples' (facial) wrinkles would still be plainly visible, even after a Botox injection. By contrast, a mobile phone with which one cannot make phone calls any more (either because the microphone is broken or it is a pre-paid phone that has run out of credit) may still serve as an 'ad hoc' flashlight.

Some inspiration from product innovation under resource constraints

Example I: Popular parlance has it that 'Necessity is the Mother of Invention.' That is, if you cannot afford it, you have to rethink it. In other words, creativity arises not despite but even because of lack of resources. In other papers and books, I discussed the case of jet propulsion (Hoegl, Gibbert & Mazursky', 2008; Gibbert &Scranton, 2009; Stokes & Gibbert, 2020, pp. 46–48), from which we draw here. There is a performance dilemma: the more power one wants from the engine, the more it heats up, until eventually some parts overheat and give in. Hoegl, Gibbert & Mazrusky (2008, pp. 1382–1383) summarise the situation as follows:

> At the end of World War II, several American teams under General Electric and several German teams under BMW and Heinkel were competing against each other in the race to resolve this dilemma … The American team had a virtual blank check for buying whatever costly raw materials it needed to create the most heat resistant alloys … By contrast, the German engineers were denied access to state-of-the-art heat resistant alloys due to funding problems and post-war disruptions of international trade. They simply could not procure the required materials and had to make do with what was at hand. The resource-constrained German team eventually resolved the performance dilemma in a simple way: by focusing on developing more efficient ways of cooling their poorer alloys, rather than developing more heat resistant alloys.

The breakthrough was achieved by a so-called 'bypass' technology that removed material from the centre of the rotor blades (which are most vulnerable due to their direct exposure to heat). The rotors were literally hollowed out, allowing air to flow through them in a virtuous circle (as jets fly faster, more air passes through the rotor blades cooling them off even more efficiently). This particular bypass technology was not at all new; in fact, it had been used for piston engines, but the resource constraints of the German team made its utility in the new context more blatantly visible.

Example II: Music without music: there is another nice example in the same category of making do with what is at hand: in 2001, iTunes sold a song that contained no music, no lyrics, essentially it was a no-song, i.e. complete silence. Does this make any sense? It does, once one 'accepts' the song without song as, well, something else (i.e. re-categorises the 'song' that no longer is a song). The main point also here is that we need to 'make do' with what we find. A song without music is essentially silence (rather than music) – 60 seconds of silence, in fact. Does this make the song any more attractive? Not until what cognitive psychologist Barsalou calls an 'ad hoc category' for this silence is found. The revenues for iTunes' one-minute silence constitute in actual fact a donation to help the people of Haiti, struck by a major natural disaster in 2001. Caritas launched this idea together with Universal Music, Switzerland and the advertising agency Jung von Matt/Limmat. The idea of selling a song without music is actually not a new idea. In the 1950s jukeboxes became popular. So popular, in fact, that Columbia saw the need to press a three-minute 'no-song' song on their LPs so that people in a diner could have at least some moments without the jukebox playing tunes in the background all the time. In fact, this 'no-song' turned out to be one of their most played songs (www.stumblerz.com/silence-is-it-gold/). Crucially, though, this absence of a song, or silence, was just that: silence. Unlike in the iTunes-Caritas case, it was not re-categorised to become something other than silence. In fact, it took 60 years, a massive earthquake, and the creative energies of Caritas, a major music label, and one of the world's most innovative advertising agencies to re-categorise the absence of a song where a song should be, and to make consumers pay for it.

Example III: Washing machines. Sull and Ruelas Gossi (2004) describe how Haier, the largest Chinese appliance manufacturer, realised that in rural parts of China, Haier repairmen had to fix washing machines which were completely clogged up with dirt and grit and most even had stopped heating up the water. 'Well', they must have thought initially, farmers clothes are muddy and filthy so it is quite natural for the grit to end up in the washing machine's system, right? But quite *so* much dirt?! After some time, the repairmen discovered that small-scale farmers actually 'recycled' the washing machines which no longer heated the water as a quick and efficient way of cleaning vegetables for sale on the local market. Now – how would you have responded to this discovery? By chastising your customers to not abuse washing machines for tasks they were clearly not intended? By cancelling warranty in case of washing machine abuse? Or by launching an innovative new product specifically designed to launder veggies, rather than pants? Or perhaps by jumping off the path of least resistance and actually seeing this as an opportunity to re-categorise the very product category? Haier chose the latter. The result was putting a special, 'vegetable' programme button on their machines, and installing easier-to-clean filters so that people could wash both their clothes as well as their vegetables. Even a 'washing' machine with a special programme for making goats' cheese was developed in this way (Sull & Ruelas Gossi, 2004, p. 11).

Example IV. Water purification. It sometimes seems remarkable just how 'close' at hand a new product idea really is when you are faced with a product that no longer works. Consider water purification. Many a team has struggled with the task of developing small-scale water purifiers that can be used in the developing world, or affordable tablets that somehow sterilise water. Intuitive solutions probably include pouring the water through some sort of filter or putting some sort of tablet into the water. Both, however, tend to cost more money than the average shack resident may be able (or willing to) afford in the long run. A team of Swiss scientists found that water can be purified by putting it in a plastic bottle, leaving it on a corrugated iron roof and leaving it in the sun for six hours. The heat and ultra-violet rays of the sun kill all the germs. Ironically, it is precisely the (now dysfunctional) tool that 'is at hand' of a thirsty person (an empty water bottle), that holds the potential solution for quenching thirst sustainably and healthily.

Example V. Two scooters and a fridge make a car (as discussed in Hoegl, Gibbert & Välikangas, 2009, pp. 14–18): an opportunity for product innovation management in diversified enterprises lies in taking advantage of bricolage to creatively combine existing resources that are no longer used. Consider an extremely fuel-efficient micro car built more than half a century ago, BMW's Isetta (it consumed only 2.7 litre/100 km, or 87 mpg). The origins of this car were with the Italian firm Iso SpA. According to the popular myth, in the early 1950s, the company was building refrigerators, motor scooters and small three-wheeled trucks. Iso's owner, Renzo Rivolta, decided he would like to build a small car for mass consumption. By 1952 the engineers Ermenegildo Preti and Pierluigi Raggi had designed a small car that used the engine of discontinued line of scooters and named it Isetta – an Italian diminutive meaning little Iso. It is said that the two engineers (or should we say, *bricoleurs*) had arrived at the award-winning design of the Isetta by taking two scooters, placing them side-by-side and adding a refrigerator door to enter the contraption.

Overall, then, the two factors underlying bricolage, ad hoc categorisation and use innovation may be instrumental in pushing many an *ingénieur* off his or her path of least resistance, providing a new impetus in both the practice and pedagogy of teaching sustainability. As a university lecturer, we would like to encourage you, with this book, to think instead as a *bricoleur*, to use resource constraints (especially those arising from pedagogical approaches that are dysfunctional) and to be inspired by some of the examples here to further innovate and extend the wicked pedagogy we propose here.

References

Akrivou, K. & Bradbury-Huang, H. (2015). Educating integrated catalysts: Transforming business schools toward ethics and sustainability. *Academy of Management Learning & Education*, 14(2), 222–240.

Albernathy, W.J. (1978). *The productivity dilemma: Roadblock innovation in the automobile industry*. Baltimore: Johns Hopkins University Press.

Antonacopoulou, E.P. (2010). Making the business school more 'critical': Reflexive critique based on phronesis as a foundation for impact. *British Journal of Management*, 21, 6–25.

Antonacopoulou, E.P. & Bento, R.F. (2018). From laurels to learners: Leadership with virtue. *Journal of Management Development*, 37(8), 624–633.

Aragon-Correa, J.A., Marcus, A.A., Rivera, J.E. & Kenworthy, A.L. (2017). Sustainability management teaching resources and the challenge of balancing planet, people, and profits. *Academy of Management Learning & Education*, 16(3), 469–483.

Baker, T. & Nelson, R.N. (2005). Creating something from nothing: Resource construction through entrepreneurial bricolage. *Administrative Science Quarterly*, 50, 329–366.

Barsalou, L.W. (1982). Context-dependent and context-independent information in concepts. *Memory & Cognition*, 10, 82–93.

Barsalou, L.W. (1983). Ad hoc categories. *Memory & Cognition*, 11, 211–227.

Bauman, Z. (2000). *Liquid modernity*. Cambridge: Polity Press.

Bower, J.L. & Christensen, C.M. (1995). Disruptive technologies: Catching the wave. *Long Range Planning*, 28(2), 155. doi:10.1016/0024-6301(95)91075-1

Brown, C.L. & Carpenter, G.S. (2000). Why is trivial important: A reasons based account for the effect of trivial attributes on choice. *Journal of Consumer Research*, 26(4), 372–384

Christensen, C.M. (1997). *The innovators dilemma: When new technologies cause great firms to fail*. Boston, MA: Harvard Business Review Press.

Christensen, C.M. & Raynor, M.E. (2003). *The innovators solution: Creating and sustaining successful growth*. Boston, MA: Harvard Business School Press.

Christensen, C.M., Anthony, S.D. & Roth, E.A. (2004). *Seeing what's next: Using the theories of innovation to predict industry change.* Boston, MA: Harvard Business School Press.

Christensen, C.M., Anthony, S.D., Berstell, G., Nitterhouse, D. (2007). Finding the right job for your product. *MIT Sloan Management Review,* 48(3), 38–47.

Dahl, D.W., & Moreau, C.P. (2007). Thinking inside the box: Why consumers enjoy constrained creative experiences. *Journal of Marketing Research,* 44(3), 357–369.

Duymedjian, R. & Rüling, C. (2010). Towards a foundation of bricolage in organization and management theory. *Organization Studies,* 31(2), 133–151.

Estes, Z., Guest, D., Gibbert, M. & Mazursky, D. (2012). A dual-process model of brand extension: Taxonomic, feature-based and thematic, relation-based similarity independently drive brand extension evaluation. *Journal of Consumer Psychology,* 22(1), 86–101.

Farjoun, M. & Lai, L. (1997). Similarity judgments in strategy formulation: Role, process, and implications. *Strategic Management Journal,* 18(4), 255–273.

Ferraro, F., Etzion, D. & Gehman, J. (2015). Tackling grand challenges pragmatically: Robust action revisited. *Organization Studies,* 36(3), 363–390.

Finke, R.A., Wardt, T.B. & Smith, S.M. (1992). *Creative cognition.* Cambridge, MA: MIT Press.

Garud, R. & Karnoe, P. (2003). Bricolage versus breakthrough: Distributed and embedded agency in technology entrepreneurship. *Research Policy,* 32, 277–300.

Gentner, D. (1982). Structure mapping: A theoretical framework for analogy. *Cognitive Science,* 7, 155–170.

Gentner, D. & Markman, A.B. (1997). Structure mapping in analogy and similarity. *American Psychologist,* 52, 45–56.

Gibbert, M. & Mazursky, D. (2007). A recipe for creating new products. *Wall Street Journal,* 26 October.

Gibbert, M. & Mazursky, D. (2009). How successful would a phone-pillow be? Using dual process theory to predict the success of hybrids involving dissimilar products. *Journal of Consumer Psychology,* 19(4), 652–660.

Gibbert, M. & Scranton, P. (2009). Constraints as sources of radical innovation? Insights from jet propulsion development. *Management & Organizational History,* 4(4), 1–15.

Gibbert, M., Hampton, J., Estes, Z. & Mazursky, D. (2012). The curious case of the fridge-TV: Dissimilarity and hybridization. *Cognitive Science,* 36(6), 992–1018.

Gibbert, M., Hoegl, M. & Välikangas, L. (2014). Introduction to the special issue: Financial resource constraints and innovation. *Journal of Product Innovation Management,* 32(2) 197–201.

Gill, T. (2008). Convergent products: What functionalities add more value to the base? *Journal of Marketing,* 72, 46–62.

Gill, T. & Dube, L. (2007). What is a leather iron or a bird phone? Using conceptual combinations to generate and understand new product concepts. *Journal of Consumer Psychology,* 17(3), 202–217.

Gill, T. & Lei, J. (2009). Convergence in the high-technology consumer markets: Not all brands gain equally from adding new functionalities to products. *Marketing Letters,* 20, 91–103.

Goldenberg, J. & Mazursky, D. (2002). *Creativity in product innovation.* Cambridge: Cambridge University Press.

Goldenberg, J., Lehmann, R.D. & Mazursky, D. (2001). The idea itself and the circumstances of its emergence as a predictor of new product success. *Management Science,* 47(1), 69–84.

Green, P.E. & Srinivasan, V. (1978). Conjoint analysis: Issues and outlook. *Journal of Marketing,* 4, 103–123.

Green, P.E. & Srinivasan, V. (1990). Conjoint analysis: New directions with implications for research and practice. *Journal of Marketing*, 54(4), 3–19.

Gregan-Paxton, J., Hoeffler, S. & Min, Z. (2005). When categorization is ambiguous: Factors that facilitate the use of a multiple category inference strategy. *Journal of Consumer Psychology*, 15(2), 127–140.

Hardy, C. & Tolhurst, D. (2014). Epistemological beliefs and cultural diversity matters in management education and learning: A critical review and future directions. *Academy of Management Learning & Education*, 13(2), 265–289.

Henderson, R.M. & Clark, K.B. (1990). Architectural innovation: The reconfiguration of existing product technologies and the failure of established firms. *Administrative Science Quarterly*, 35(1), 9–30.

Herstatt, C. and von Hippel, E. (1992). Developing new product concepts via the lead user method: A case study in a 'low- tech' field. *Journal of Product Innovation Management*, 9(3), 213–221.

Hirschman, E. (1980). Innovativeness, novelty seeking, and consumer creativity. *Journal of Consumer Research*, 7, 283–295.

Hoegl, M., Gibbert, M. & Mazursky, D. (2008). Financial constrains in innovation projects: When is less more? *Research Policy*, 37, 1382–1391.

Hoegl, M., Weiss, M., Gibbert, M. & Välikangas, L. (2009). Strategies for breakthrough innovation. *Leader to Leader*, 2009(54), 13–19.

Kristensson, P., Gustafsson, A. & Archer, T. (2004). Harnessing the creative potential among users. *Journal of Product Innovation Management*, 21(1), 4–14.

Leonard-Barton, D. (1981). Voluntary simplicity lifestyles and energy conservation. *Journal of Consumer Research*, 8, 243–251.

Lévi-Strauss, C. (1966). *The savage mind*. Chicago: University of Chicago Press.

Markides, C. (2006). Disruptive innovation: In need of better theory. *Journal of Product Innovation Management*, 23, 19–25.

Maślikowska, M. & Gibbert, M. (2019). The relationship between working spaces and organizational cultures. *Facilities*, https://doi.org/10.1108/F-06-2018-0072

Moon, Y. (2005). Break free from the product life cycle. *Harvard Business Review*, 83(5), 86–94.

Moreau, C.P., Markman, A.B. & Lehmann, D.R. (2001). 'What is it?' Categorization flexibility and consumers' responses to really new products. *Journal of Consumer Research*, 27, 489–498.

Mortensen, M. & Haas, M. (2018). Perspective – rethinking teams: From bounded membership to dynamic participation. *Organization Science*, 29(2), 191–355.

Nowlis, S.M. & Simonson, I. (1996). The effect of new product features on brand choice. *Journal of Marketing Research*, 33, 36–47.

Olson, J.C. & Thomas J.R. (1983). Understanding consumers' cognitive structures: Implications for advertising strategy. In L. Percy & A.G. Wotxlside (Eds.), *Advertising and consumer psychology* (pp. 77–90). Lexington, MA: Lexington Books.

Rajagopal, P. & Burnkrant, R.E. (2009). Consumer evaluations of hybrid products. *Journal of Consumer Research*, 36(2), 232–241.

Ridgway, N.M. & Price, L.L. (1994). Exploration in product usage: A model of use innovativeness. *Psychology & Marketing*, 11(1), 69–84.

Rittel, H.W.J. & Webber, M.M. (1973). Dilemmas in a general theory of planning. *Policy Sciences*, 4, 155–169.

Rust, R.T., Thompson, D.V. & Hamilton, R.W. (2006). Defeating feature fatigue. *Harvard Business Review*, 84, 98–107.

Schreier, M., & Pruegl, R. (2008). Extending lead user theory: Antecedents and consequences of lead userness. *Journal of Product Innovation Management*, 25, 331–346.

Sloman, S.A., Love, B.C. & Ahn, W. (1998). Feature centrality and conceptual coherence. *Cognitive Science*, 22, 189–222.

Stokes, P.D. & Gibbert, M. (2020). *Using paired constraints to solve the innovation problem.* Springer International Publishing.

Stremersch, S. & Tellis, G.J. (2002). Strategic bundling of products and prices: A new synthesis for marketing. *Journal of Marketing*, 66, 55–72.

Sull, D. & Ruelas-Gossi, A. (2004). The art of innovating on a shoestring. *Financial Times, Mastering Innovation*, 24 September, 10–11.

Thompson, D.V., Hamilton, R.W. & Rust, R.T. (2005). Feature fatigue: When product capabilities become too much of a good thing. *Journal of Marketing Research*, 44, 431–442.

Tushman, M.L. & Anderson, P. (1986). Technological discontinuities and organizational environments. *Administrative Science Quarterly*, 31(3), 439.

Upper, D. (1974). The unsuccessful self-treatment of a case of writer's block. *Journal of Applied Behavioral Analysis*, 7(3), 497.

Ward, T.B. (1994). Structured imagination: The role of category structure in exemplar generation. *Cognitive Psychology*, 27(1), 1.

Ward, T.B., Dodds, R.A., Saunders, K.N. & Sifonis, C.M. (2000). Attribute centrality and imaginative thought. *Memory & Cognition*, 28(8), 1387.

Välikangas, L. & Gibbert, M. (2015). *Strategic innovation: The definitive guide to outlier strategies.* Upper Saddle River, NJ: Financial Times Press.

Weick, K.E. (1993). The collapse of sensemaking in organizations: The Mann Gulch disaster. *Administrative Science Quarterly*, 38, 628–652.

Weiss, M., Hoegl, M. & Gibbert, M. (2011). Making virtue of necessity: The role of team climate for innovation in resource-constrained innovation projects. *Journal of Product Innovation Management*, 28(1), 196–207.

13

PROBLEMS YOU SOLVE AND PROBLEMS YOU WORK ON

Connecting to, and engaging with, society and corporations

Liisa Välikangas

Introduction

Some years back in San Francisco I had the privilege of talking with George Shultz, the former Secretary of State of the United States. Describing his long experience in global politics, he concluded: 'There are problems you solve, and problems you can only work on.' In my understanding he meant some challenges are so hard that no ready solution is feasible, so the best we can do is to keep working on them. Not to give up. This distinction between solving and tackling has kept me thinking. It is an inspiration to this chapter, the pedagogy, and the efforts around a master's course that invites students from three universities together with a global corporation to address global challenges that have been described as grand (Ferraro et al. 2015, Hilbert 1902), wicked (Rittel & Webber, 1973), or complex and messy (Ackoff, 1981). They are sorts of moonshots in aspiration but often mere propeller planes in implementation with characteristics of collective action problems (Dietz, Ostrom & Stern, 2003; Hardin, 1968; Ostrom, 1990).

Our challenge for the past three years in the TWC course between the three universities has been food waste and loss. According to the Food and Agriculture Organization of the United Nations, about one-third of food produced globally, or 1.3. billion tonnes per year, goes to waste (www.fao.org/save-food/resources/key findings/en/). While 820 million people go hungry, this waste and loss contributes significantly to CO_2 emissions, lack of water and loss of resources around the world. Food waste and loss is an important challenge should we wish to tackle well-being but also climate change. It is a grand challenge as acknowledged by UN Sustainable Development Goals but also a wicked problem in that it has many contradictory aspects – some of which manifest in the fight to save the Amazon against clearing for meat and soy production for which there is global demand. And there is no one single approach that could persuade all of us to change our behaviour today or tomorrow. Any attempted solutions may have unintended side-effects as in organic farming, which tends to produce food at a higher retail price point (even though the environmental cost is lower). Given the growing population on earth, how can we feed everyone if our environment further deteriorates? Human impact in an era called Anthropocene has already changed our natural environment irreversibly. How much further can we enlarge the human footprint while reducing the diversity of, and living conditions for other species? There is radical uncertainty in any approach, and the problem may change its nature depending on the perspective. Varying stories can be told that are reasonable yet potentially conflicting (Verweij et al., 2006). For example, is climate change due to extravagant consumption of those (too) well off or is it a matter of lack of global governance for the way markets operate (in not accounting for externalities) or perhaps it is not really a problem at all but something to be solved in due course by human ingenuity and technological development? We thus narrate ourselves to (in) action that may be quite different in its aim and scope and in its consequences.

Tackling Grand Challenges in three pillars

We need robust action to tackle grand challenges. Robust action has been defined as 'action that accomplishes short-term objectives while preserving long-term flexibility' (Eccles & Nohria, 1992, p. 11). In a seminal study of Renaissance Florence and in particular of Cosimo de Medici, its 'sphinx-like' power broker, the authors Padgett and Ansell (1993) point to ambiguity and heterogeneity as building blocks of powerful action in everyday life. Ferraro et al. (2015) later build on this notion of robust action as they develop their well-known framework for Tackling Grand Challenges, such as climate change or food waste and loss. Yet for very large-scale problems, more is needed than one person's capacity or unilaterality. No Cosimo de Medici is able to solve food waste or climate change alone!

What makes action robust in Tackling Grand Challenges, the authors suggest, is developing a participatory architecture, defined as 'a structure and rules of engagement that allow diverse and heterogeneous actors to interact constructively over prolonged timespans' (Ferraro et al. 2015, pp. 373–374). This is an organising challenge: how do we invite and engage many different actors – people, corporations, non-profits, governments and beyond – to join in the conversation about how to tackle a particular challenge? An example of such a participatory architecture is the United Nations, which was formed to develop relations and foster cooperation between nations for the cause of world peace and everyone's well-being. More recent examples may be platforms such as Ushahidi, originally formed to report post-election violence in Kenya in 2008. The platform has evolved to support the bottom-up flow of information where reports of earthquake victims and other locally gathered bits of information can be collected and visualised as a map. Anyone can send observations on the ground through a text message.

The second pillar of Tackling Grand Challenges is multivocal inscription, or the capacity to 'sustain different interpretations among various audiences with different evaluative criteria, in a manner that promotes coordination without requiring explicit consensus' (Ferraro et al. 2015, p. 375). This means not only that different perspectives are welcome but that they are sustained over time so that no particular voice becomes dominating. An example is the concept of Sustainable Development, launched by the World Commission on Environment and Development in their report Our Common Future in 1987, where environment and development were linked together as a pragmatic but also as a moral imperative. Both environmental and economic considerations matter; both voices needed to be heard. In a democracy – governance by majority – minority rights are still protected. The United Nations Declaration on Minorities, Article 27 (2015), states:

> In those States in which ethnic, religious or linguistic minorities exist, persons belonging to such minorities shall not be denied the right, in community with the other members of their group, to enjoy their own culture, to profess and practise their own religion, or to use their own language.

Multivocality is important not only as a principle of democratic human rights but as a way to tackle challenges where no one approach brings a lasting solution. A battle can be won in a show of overpowering force but peace must be sustained when living together. In such living together, multivocality requires listening to others even when in disagreement. Voicing the disagreement is also multivocality. This requires curiosity and respect for the Other together with wisdom to live in societal contexts that have no absolute truths. Of course, going back to Cosimo de Medici and the study of his effectiveness:

> The key to understanding Cosimo's sphinxlike character … we argue, is multivocality – the fact that single actions can be interpreted coherently from multiple perspectives simultaneously, the fact that single actions can be moves in many games at once, and the fact that public and private motivations cannot be parsed. (Padgett & Ansell, 1993, p. 1263)

Multivocal inscription is also potentially strategic behaviour.

Tackling Grand Challenges still requires a third pillar, which is called distributed experimentation. Such experimentation is rooted in the notion that uncertainties prevent us from knowing the best strategy; yet we cannot remain without taking any action at all. Absent a grand solution, Ferraro et al. (2015, p. 376) suggest a need for 'iterative action that generates small wins, promotes evolutionary learning, and increases engagement, while allowing unsuccessful efforts to be abandoned'. Examples abound in renewable energy where solar power, wind mills, ocean waves, geothermal heating, hydropower and other ways to reduce carbon footprint are being explored. Will we all drive electric cars soon? Is charging the car battery good for the environment or just a way to push the fossil fuel out of mind as we do not need to go to a petrol station anymore? In any case, distributed experimentation seems to be thriving across industries. There are many companies related to health and well-being that provide ways to measure one's exercise, diet, sleep, networking and so on. Nonprofit groups such as Biocurious, a hackerspace for biotech in Northern California, provides a lab for experimenting how to improve one's performance at work. Their motto is: 'Experiment with Friends!'

Connecting to, and engaging with, society

Perhaps what is missing in the framework presented above is the notion that there are actors which are more multi-powered than others. In Florence, Cosimo de Medici, a virtuoso of heterogeneity and a master of ambiguity, acted as judge and boss at the same time, something that is a contradiction and a conflict of interest. The executive and judiciary powers ought to be separate. Padgett and Ansell (1993, p. 1260) write:

> The contradiction, in state building or in any organisation, is between judge and boss: founders cannot be both at once. Stable self-regulating maintenance of rules (i.e., legitimacy) hinges on contending actors' conviction that judges and rules are not motivated by self-interest (Elster 1983; Padgett 1986; Douglas 1986). At the same time, the nightmare of all founders is that their organisational creation will walk away from them. As Weber recognised long ago, in crisis (sooner or later inevitable), direct intervention in or overt domination of locked-in interactions is a sure sign of control's absence, not of its presence. Tactical tinkering to maintain fleeting control sucks in founders to locked-in role frames, thereby inducing attributions of self-interest and undermining their judicial perch above the fray.

Thus in pursuing grand challenges pragmatically, powerful or institutional actors do matter. When working with Universitá Svizzera Italiana's Middle Eastern Mediterranean Platform, this quickly becomes evident. Platforms where people share thoughts openly have to protect the identity of the person for fear of repercussions. At the same time, such platforms would be highly effective if they were able to collect instances of corruption and present a more complete picture. Where is corruption actually happening? Can it be documented?

Institutions matter too. Participatory architectures present a hopeful picture of everyone contributing yet sometimes individual action is not sufficient without a working state or a rule-oriented society. Failed states present extreme conditions where grand challenges abound, yet approaches seem to require a peace agreement first in case of warring parties and enough security for civil action. In Afghanistan, a non-profit Innovation Democracy (co-founded by this author) taught many students of the Kabul University Economics Department skills such as team work, creativity and business planning that helped them find work after graduation. Such activity would have been impossible or too dangerous if the university had not allowed the local Afghan (woman) teacher (with a bachelor's degree earned abroad) to offer the course. Luckily the university (mostly) supported the course even though difficulties occasionally occurred with other authorities.

Another consideration is resilience. Will the actions – starting with experimentation – have staying power? Will the experiments actually evolve into further refined activities based on learning in the process of experimentation? One might also ask how will the experiments be scaled up, or gain impact? Perhaps from the perspective of resilience a better question to ask is how the experiments compound: how is the variety, the multivocality, preserved as we learn more about the impact of the experimentation? There is a risk that one experiment producing the best outcomes (in some sense) or the hoped for results at a particular moment in time, will be copied and scaled up at the expense of other experiments and learnings in the future. This might exclude any further learning, thus reducing strategic resilience. Better to maintain a multitude of experiments and combine their effects: to compound rather than scale change; to have many approaches working together rather one single approach used everywhere. For example, instead of everyone focusing on artificial intelligence or its underlying disciplines (mathematics, coding) as many call for in industry today, perhaps it would be better to study a variety of disciplines while gaining understanding of the basics of machine learning, its opportunities and dangers of biased data and algorithmic operation. At MIT, a new college will combine AI, machine learning and data science with other academic disciplines (Technology Review, 15 October 2018). The intent is to teach all students these skills. To the extent that other academic disciplines are equally honoured, this might increase the resilience of students' professional skills.

In a website linked to the book *Future and Its Enemies* (Postrel, 1998), Virginia Postrel writes persuasively about such resilient capability for changing, or the importance of conducting experiments in a distributed, decentralised way:

> unplanned, open-ended trial and error – not conformity to one central vision – is the key to human betterment. Thus, the true enemies of humanity's future are those who insist on prescribing outcomes in advance, circumventing the process of competition and experiment in favor of their own preconceptions and prejudices.

There is danger that experimentation which reaches an outcome envisioned at the time will prevent us from finding what we really are capable of as people and societies, as it was not imagined at the time. Our capacity to imagine may then limit us. However, our imaginations develop, and our aspirations change. What seems highly desirable may no longer seem so once achieved. Perhaps we want more, or we want something entirely different, hitherto unimagined. Or we find out what we wanted is not worth it: e.g. economic prosperity at the expense of natural environment. Aspirations are known to adapt to experience, hopefully fast enough to save life and its supporting environment.

Working with a global company

One avenue for gaining traction in Tackling Grand Challenges is working with a global corporation. Such collaboration is an art in itself, requiring an understanding of the business vocabulary, a sensitivity to the corporate way of working, and an ability to connect and engage around an issue of importance to the company and its leadership. Large companies often move slowly at first but may eventually have a large impact through their measured actions. Patience may be needed but the ability to create urgency and relevance around the challenge is equally important. How to get someone's attention is one of the basic yet demanding skills. Planning and calendaring ahead of time is important. How to maintain the momentum of collaboration so it is not left at the end of a long list of things to do in the company. Everyone is busy; why should they take yet another thing on? How do I communicate the point in a few sentences? How do I ignite excitement? Why should the company, and the person I will be talking with, be interested in my proposal? Maintaining good personal relationships is important for working together.

Step 1: Developing a participatory architecture

When taking the three pillars of Tackling Grand Challenges and applying them to working with a large company, the first task is to establish the participatory architecture for the collaborative effort. It is important to have a point person in charge of the collaboration in the company. Equally important is to have someone, a faculty member, in charge of the collaboration on the side of the course. As students work in teams, each team should appoint a person who is responsible for the company relationship – communicating with them and sharing information among the student team as well. This student represents the team to the company. The participatory architecture should also be translated into regular meetings. What is the pacing of the work and how does it manifest in conference calls, company visits or other presentations and exchanges? It is recommended that at least one communication, often a conference call, happens every second week so that the project does not lose its momentum. Such calls should be scheduled at least a month ahead of time as calendars are usually full.

In addition to the communication protocol, the collaboration often involves working with the company on site. For example, with IKEA cafeterias, student teams engaged in various local experiments exploring food waste and loss behaviour. This required its own participatory architecture with people as contact points in the particular cafeterias where the experiments were implemented. This kind of grassroots work can be challenging as it often requires the busy local staff to learn and appreciate the point of the experimentation. It is important to share the motivation of the work and its learnings with everyone involved and also keep the company participants posted about the progress. With Nestlé, student teams worked with different local operations ranging from coffee brewing to tomato growing and to baby food production with different research agendas. It is important to regularly communicate what is happening in the next two weeks. What have we done so far? Any interesting lessons learnt to discuss? What questions do we have? Discussing the results with local staff can increase insight tremendously as there may be contextual issues that are otherwise lost or ignored.

Step 2: Maintaining multivocality

Multivocality may sound like a value more attuned to democracy than corporations, yet it is still important. In its simplicity, multivocality suggests opening the collaboration to many different parties within the corporation – not only the leadership but also employees or middle managers. Open innovation has gained traction in making this point that not all good ideas come from inside a particular privileged group, or even inside the company, but often can be found far away from the corporate centre, maybe in the market place or even far away geographically. Thus harnessing ideas broadly is important for their quality. Similarly, it would be desirable to have diversity in collaborative participation. In part this diversity is provided by the students who come from different backgrounds, nationalities, educational discipline, age and so on. Yet it is also important to have the company broadly represented for multivocality.

The second aspect of multivocality goes beyond the diversity of participants. It is about sustaining multiple perspectives to the issue as it is explored. These perspectives can be a sort of narrative about what is happening and what the consequences might be – much like hypotheses for experimentation. But the multivocal inscription may also genuinely represent different motivations, interests and priorities that are important to keep in mind when tackling the problem. Do we want a solution that is brought by private companies? Do we think it is the government's role and responsibility to act legislatively? Do we privilege the poor or the highly skilled, and so on. An example is education in Scandinavian countries, which is free. This has made universities accessible to anyone independently of financial resources. Yet some argue that paying for education would motivate students to study harder, or at least faster. Others claim this would lead to a loss of equality, an important Nordic value. The question is complex. It also includes a university's perceived role – market-based or public service? James G. March (2003, p. 206) in his article 'A Scholar's Quest', wrote:

> A university is only incidentally a market. It is more essentially a temple —a temple dedicated to knowledge and a human spirit of inquiry. It is a place where learning and scholarship are revered, not primarily for what they contribute to personal or social well-being but for the vision of humanity that they symbolise, sustain, and pass on.

This is a call for multivocality in its poetry. In working with corporations, such a call may be translated as a commitment to Tackling Grand Challenges in a way that sustains a vision of humanity that is based on a joint future. In *Down to Earth* (2018), Bruno Latour writes about the dangers of such a commitment to a joint future fading. We may no longer believe that there is a future that is shared among all people.

Step 3: Engaging in distributed experimentation

Faced with uncertainty, experimentation is a good way to proceed. Karl Popper is said to have noted that experimentation allows that our hypotheses die in our stead. It is thus a powerful method for learning with low risk or less risk. Experimentation has become the norm in lean start-up thinking where minimally viable (experimental) products are launched on the market place to learn about customers' reactions. This was presented by the start-up gurus as an alternative to planning. The motto is Fail Fast with an understanding that each failure brings some new knowledge about how to develop the product or offering further. It also gains time by cutting planning. Experimentation allows us to learn from a broader experience base than what we would usually be exposed to by entering arenas where we would otherwise not be present. Experimentation is a way to enter lightly, in an exploratory mode, and it should happen with a hypothesis as to what we are trying to learn. The commitment is to the learning not to a particular hypothesis. What do we expect to happen and why? What is our theory as to why the experiment should create this result?

Experimentation should also be distributed. This means that it should take place in different arenas of uncertainty, perhaps in different markets, in different formats. Perhaps entirely different experiments will be run. Working with IKEA, distributed experimentation was evident in working with a number of cafeterias in different Scandinavian cities. With Nestlé, the work was with factories focused on different supply chains. Important here – if once wishes to be more strategic – is to think about what are the major uncertainties that experimentation is seeking to tackle and then design experiments to address those. For example, in food waste is the most demanding issue to be found inside the cafeteria or factory? Is it in the farm? Or the consumer's home? How might we approach supply chains so that waste and loss became more visible? What is our hypothesis about why highly efficient supply chains still have food loss at all as wasting is costly? Why does the market discipline not eliminate food waste and loss which is costing money and resources? Understanding these kinds of issues, it is likely that iterative experimentation (and studying the issue) is needed – once an experiment is complete, it gives an understanding of the next questions to ask in the pursuit of a more holistic view.

Senior executives are often worried about too much experimentation leading to the loss of focus in the main business. Intel's former CEO Andy Grove famously used to call such focus 'Job 1'. There are ways to make sense of experimentation in a more strategic way to guide it. For example, if the company has a notion of its main strategic pursuits – such as mobility, machine intelligence, aesthetics –

experiments can provide information about these change directions or vectors. Experiments that add to the understanding of mobility, as an example, are encouraged and iterated, whereas those that are in some other less relevant arena can be discouraged and ended. Nevertheless, experimentation may convey important local information once people in a company use it as a strategic tool for learning.

In working with large companies, some further advice may be useful. Often the grand challenges are very large issues by definition and may benefit from more oblique approaches than direct head-on attacks. John Kay from the London School of Economics has written about obliquity, or the indirect pursuit of goals and objectives: 'Strange as it may seem, overcoming geographic obstacles, winning decisive battles or meeting global business targets are the type of goals often best achieved when pursued indirectly. This is the idea of Obliquity' (www.johnkay.com/2004/01/17/obliquity/). Obliquity is particularly recommended when there is a lot of uncertainty and the outcomes depend on interactions with other people, something grand challenges inevitably do. An example of a direct approach might be found in a dark thriller by Dan Brown called *Origin*. How does artificial intelligence create world peace? By eliminating the human race. Straight to the point but not very desirable from the human perspective! An indirect approach would require interactions with people, trust building, conversations, respecting the Other, political and legal agreements, and much more. The European Union's predecessors, the European Coal and Steel Community (1951) and the European Economic Community (1957), were founded to create economic ties among European nations and hence – obliquely – foster peace. This also suggests the solution ideas may come detached from the problem – end of war but then economic exchange. Similarly, fuel for airlines is tax free when the plane arrives at an international destination according to an agreement made in 1944 (the Chicago Convention). It was considered important for people to get to know each other after the Second World War to prevent another war. The unforeseen consequences include the burden air travel now poses for the climate.

Conclusion

One should not forget, in the midst of experimentation, that Tackling Grand Challenges, or wicked problems, requires innovation, 'the pursuit of bold ideas and the adoption of less conventional approaches to tackling large, unresolved problems' (Colquitt & George, 2011, p. 432). Of course, such bold ideas may entirely fail or may take us to a situation even worse than what we started with. Yet I read this call for boldness as an invitation for courage. It is likely that Tackling Grand Challenges requires a lot of courage in being unconventional, questioning established orthodoxy and presenting ideas that are potentially novel. It is important, however, to build a participatory architecture around these ideas and their experimentation and maintain the multivocality that is necessary for the next big bold idea to surface.

References

Ackoff, R.L. (1981). On the use of models in corporate planning. *Strategic Management Journal*, 2, 353–359.

Colquitt, J.A. & George, G. (2011). Publishing in AMJ: Topic choice. *Academy of Management Journal*, 54, 432–435.

Dietz, T., Ostrom, E. & Stern, P.C. (2003). The struggle to govern the commons. *Science*, 302, 1907–1912.

Eccles, R.G. & Nohria, N. (1992). *Beyond the hype: Rediscovering the essence of management*. Boston, MA: Harvard Business School Press.

Ferraro, F., Etzion, D. & Gehman, J. (2015). Tackling grand challenges pragmatically: Robust action revisited, *Organization Studies*, 36(3) 363–390.

Hardin, G. (1968). The tragedy of the commons. *Science*, 162, 1243–1248.

Hilbert, D. (1902). Mathematical problems. *Bulletin of the American Mathematical Society*, 8 (10), 437–479.

Kay, J. (2011). *Obliquity: Why our goals are best achieved indirectly*. London: Penguin Press.

Latour, B. (2018). *Down to earth: Politics in the new climatic regime publisher*. Cambridge: Polity Press.

March, J.G. (2003). The scholar's quest. *Journal of Management Inquiry*, 12, 205–207.

Ostrom, E. (1990). *Governing the commons: The evolution of institutions for collective action*. New York: Cambridge University Press.

Padgett, J.F. & Ansell, C.K. (1993). Robust action and the rise of the Medici, 1400–1434. *American Journal of Sociology*, 98, 1259–1319.

Postrel, V. (1998). *The future and its enemies: The growing conflict over creativity, enterprise, and progress*. New York: Free Press.

Rittel, H.W.J. & Webber, M.M. (1973). Dilemmas in a general theory of planning. *Policy Sciences*, 4, 155–169.

Technology Review. (2018, 15 October). The smartphone app that can tell you're depressed before you know it yourself. Retrieved from https://www.technologyreview.com/2018/10/15/66443/the-smartphone-app-that-can-tell-youre-dep ressed-before-you-know-it-yourself/

Verweij, M., Douglas, M., Ellis, R., Engel, C., Hendriks, F., Lohmann, S. & Thompson, M. (2006). Clumsy solutions for a complex world: The case of climate change. *Public Administration*, 84, 817–843.

PART IV

Feedback: Enjoying, sharing, dinner, partnering, scaling, entrepreneurial

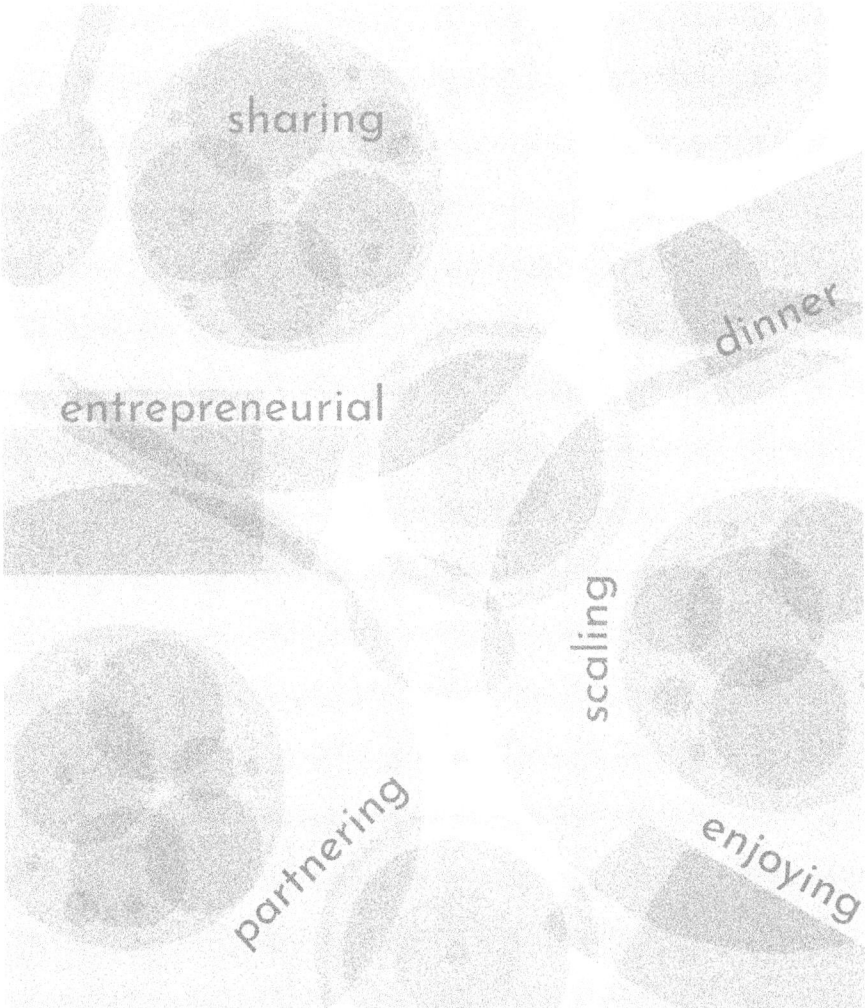

sharing

dinner

entrepreneurial

scaling

partnering

enjoying

Drawing by Inês Costa

14

SHARING STUDENT VOICES

Monika Maślikowska

Introduction

Students of the 'World in the Making: Tackling World Challenges' course worked together for about three months in diverse globally distributed virtual teams composed of members with different backgrounds in terms of universities, education, nationalities, spoken languages, gender and age. To better understand the student perspective of the experience in the project, we decided to capture numerous observations (through notes and photographs), ask for direct student feedback through the pre- and post-course surveys and conduct interviews (recorded and transcribed) with course participants.

Students completed surveys with open- and closed-ended questions, evaluating and describing their intercultural and virtual team skills, their views on food waste, and their overall Wicked Learning experience. After the completion of the course, 61% of students stated that the course changed their views on working in an intercultural environment, about 30% of students stated that they improved their virtual collaboration skills, and 78% of students indicated that they developed an understanding of how to approach wicked problems.

In order to gather more specific and detailed data on student views, we conducted a series of semi-structured in-depth interviews, which took place in the beginning and at the end of the course, in various settings – on campus, at the IKEA store, at the airport, in the plane, on the street, in the boat – to assure a casual atmosphere while capturing current experiences. Interviews revealed many nuances regarding team member diversity, working styles, leadership, communication patterns, physical working spaces in remote work, as well as the general factors affecting team collaboration.

Despite the general challenges associated with diversity, uncertainty associated with wicked problems, and physical distance, the overall feedback from students indicates that the applied pedagogy allowed for 'tackling wicked problems' in ways which would not have been possible otherwise. The sections presented below are a summary of several responses revealing an unusual view on the entire Wicked Learning experience.

Collective student voices: intercultural environment, virtual teams, wicked problems and the overall course experience

In the post-course survey, we asked students about changes in their views on working in an intercultural environment, characterisation of the virtual team experience, a developed understanding on how to approach wicked problems and the general feedback on the course. Below are questions with summaries of answers that represent the diverse student responses.

Has this class changed your views on working in an intercultural environment? If so, briefly describe how your views have changed.

While a majority of students already had previous international experiences, for some it was the first time working in such an environment. Nevertheless, in both cases, students acknowledged the significance of the exposure to different cultural backgrounds referring to nationalities or countries of origin, but also universities (often mentioning the 'university cultures'), study backgrounds, individual person-alities and ways of working. Student W said, 'the course made me further appreciate different working styles and approaches to solving problems'. While some course participants observed certain cultural differences related to national-ities, others did not. Despite different backgrounds, many students found great partners for collaboration due to similar working styles and even developed friendships on the personal level. At the same time, participants M and L indicated that, 'misunderstandings are more likely to happen when the language used during team activities is not the native language of some or all of the members'. Never-theless, the course provided a sample of potential future real-life working situations where the differences among team members present a great opportunity to include diversity of ideas. 'The multitude of perspectives and views allows for more fruitful results' (Student R) as 'multicultural teams develop innovative ideas and allow them to thrive' (Student R).

How would you characterise your virtual team experience in this class?

The general feedback provided by students indicates that the virtual team experience, despite some challenges, allowed for teamwork, which would not have been possible otherwise. As Student F explained, 'it's quite clear that working face to face has a value that is impossible to overcome, however, working remotely has some upsides (e.g. let workers take their time to work on the task … within deadlines of course)'. It is important to add that such an international project might have not been possible to achieve without the virtual collaboration. Although Student S also mentioned 'the virtual team was more effective (than face to face) in the sense of being more productive and to-the-point during the meeting, however I believe we were satisfied too quickly on solutions and therefore the output was somewhat more simplistic'. Some course participants also mentioned difficulties they encountered which were related to the different schedules among team members (who participated in other activities at the same time), technological issues, and challenging communication with the corporate partner without the possibility of meeting face to face. Several participants also referred to the challenge of maintaining motivation and engagement as stated by Student E, 'when you are working online and mainly by yourself at all times it is even more important to find enrichments that motivate and inspire everyone to take responsibility, feel that they are important and have clear tasks'. Student P added, 'Virtual teamwork benefited from having met in person at least once before starting the whole project.' In order to overcome the frequent challenges of working in virtual teams, beyond the tools already in place, such as technological training and team-building activities at the kick-off, we decided to implement a class on working in virtual teams in future course editions.

Have you developed an understanding on how to approach 'wicked problems' (problems that cannot be solved)? Please explain.

Many course participants highlighted the complex nature of wicked problems, indicating that in order to address them it is crucial to be proactive and use tools, such as academic literature and education, but also the diversity of ideas related with different (e.g. academic or cultural) backgrounds of the collaborators and the approach of taking small steps and moving forward.
Student T elaborates,

> Although I was somehow aware of different wicked problems, the course triggered my intention to be part of finding solutions. The reason for this is that I learnt to understand that one does not (and cannot) grasp the entire problem but even tackling a small part of it might lead to positive change.

Student O noted the role of education;

> The program has been a very good opportunity to approach so-called wicked problems. It is clear that these kinds of problems are impossible to solve in the short-term and with little resources. Nevertheless, it is crucial to start tackling the challenge in whatever way available to completely change your mind towards solutions. At the end of the day, it's a matter of 'try and take notes'. Finally, it has popped up that education is extremely important when it comes to wicked problems.

Diversity is important too, as Student D said, 'I really enjoyed working with students from a different academic background than my own. It made me see the problem in ways I hadn't thought of.'

Student B pointed out that understanding how to approach wicked problems 'was strongly shaped by managerial and business aspects' and that it is 'especially interesting to approach such a complex issue from the perspective of a global player (IKEA). People are used to thinking about these problems, but on an individual basis! Trying to nudge, change values etc. was awesome.'

Have your say ...

At the end of the survey, we asked the students to provide overall feedback on the entire course experience. Despite the general positive response from students, including specific lessons learned, we tried to understand key drivers for student learning and areas for improvement in the course and the applied pedagogy.

Some students referred to the importance of including consultations on virtual teamwork and the involvement of 'teachers' in the students' progress, others commented on the challenging aspect of the coordination between partner universities and the corporate partner, as well as the need for consistency in the use of the virtual communication tools. There were a number of responses that stressed the significance of clear and timely communication of course content, structure and evaluation, as well as tasks, while others felt that the flexibility that resulted from the somewhat ad hoc structure was a main advantage of the course, supporting participant engagement and empowerment in the process. Several students also mentioned that recruiting motivated students and stimulating motivation and engagement is crucial and an increased amount of team-building activities at the kick-off would be very helpful (perhaps replacing 'intercultural training', which might instead enforce stereotypes). We analysed all comments and decided to implement relevant improvements in the next course editions. Nevertheless, the general feedback was rather positive and involved some profound insights from students on the main takeaways from the course. Below there are a few indicative quotes from the student responses:

> First of all, I would like to thank you for organising of this course, you have made a great job with all coordination etc. I am also really happy that I had the opportunity to work with IKEA and also create new relationships with the other students ... it has been fun to work with this project! I have learned a lot from myself and about teamwork. It has been fun to work with IKEA and really implement an idea, usually you just come up with something but it never gets real. I have also learned a lot of new things about food waste and food security, tips on how I can minimise food waste in my daily life ... (Student I)

This was the best course I have ever had at university! There were certainly many challenges, and at times it would have been better to have more team members in the same place due to manpower needed at the location of the idea implementation. I would say an area that could be improved is communication, as at times it was unclear what we needed to do and what to expect. It may have been interesting to be able to ideate a more complex, long-term idea as I feel that we could have tackled the roots of the food waste at IKEA if there had been less restrictions of time and of having to implement it ourselves. However, overall a great experience that I would recommend to other students. (Student U)

I think the most giving learning experience of this course comes from the co-operation between the different people/universities/teams and learning how to deal with all of what that entails, rather than from the course's chosen academic/business content. (Student L)

I'd like to mention that the virtual teamwork benefited from having met in person at least once before starting the whole project, thanks for the whole organisation and fun activities! Super interesting topic and collaboration partner. One thing I think could be improved is the organisation of the course and the virtual lectures. (Student V)

Thank you – A great opportunity to work in international teams and to start making an impact! Amazing hosts in Stockholm and Lugano. THANK YOU for the side activities and food! Great course for skill set improvement: communication, team building, time management, creativity, problem-solving etc. (Student A)

Individual feedback: in-depth interviews with students

During the after-the-course interviews, students provided in-depth insights on team member diversity, working styles, leadership, communication patterns, physical working spaces in the remote work, as well as general factors affecting team collaboration. Responses involved comments on leadership types and complementarity of roles, team dynamics (e.g. team cohesion), challenges and advantages related to the diversity of team members or their working styles. Students also mentioned that the freedom of choosing a physical working space and time helps to adjust busy schedules of the team members, thus opening doors for inclusive collaboration (e.g. for students participating in multiple activities or projects at the same time). Some interviewees indicated that the practical experience of working in real-life working conditions, combined with entertainment supported student engagement in the project 'I almost would lose myself in the coursework or material, rather than focusing on other work that needed to be done, because it was so much fun and my group was fun. When people are getting along it makes the work a lot more entertaining and productive' (Interviewee 1).

The content presented below includes excerpts of several interviews with representative views of the students.

Interview 1

Q: I'm trying to investigate what were the interesting aspects of your team development in the diversity context, not only gender diversity, but also nationality and different backgrounds and the fact that the students were from different academic faculties and how that affected your work.

A: In general, I thought our team was really coherent, although you never know what you are going to get with this type of situations, especially working virtually. I would say we had two team leaders that naturally emerged due to their personality traits ... I think that was just a matter of stronger personality and really being interested in the topic and the fact that we were able to work hands on with the firm, not just reading literature and reviews, so felt it was something that could make a difference and that we could possibly see in the future at IKEA. That is what was understood by us, so we kind of took the lead. The other group members were fantastic as well, though ... we needed supporting roles to function. So, we kind of took the lead on making the museum and turning the idea into life and putting the energy behind it but then when I was going to be the one initiating all of the events, organisation and being there, (the team) created all of the prerequisites. They created all of the material that I would need, so all I had to do was to go and put that into action, so it was very fluid. Then, all of the other parts were really split up pretty evenly and they took up the lead on creating the paper. We always rotated on who was leading the meeting, so everyone also got their own leading role and being in charge of the day and setting it up with everyone helped to get a feeling of what is it like to be a leader and people just followed in line. Which was really fun, because when you are the leader, you are like: ok, here are the dates I'm available based on what you said and you were just chiming in when they asked you to during the meeting to supply information.

Q: And what about the styles of work, have you noticed some differences? And if so, why do you think they were there?

A: I think with the (indicated nationality) guy there was a bit of difference in the style of work just because

Q: Like what?

A: Way more relaxed and way less intrusive. Kind of being like: Yeah, ok I'll do that, but not like: Yes, I want this part, being more decisive. More like: If you assign something to me, I'm going to get it done in my time, but not stand up for what exactly you want to be involved in.

Q: Do you think it was culture related?

A: Partially culture, but partially maybe that he was the only male, so he kind of, was maybe like: Ok, so we have all these females, so let's let them take over and I'll do whatever. And he was super encouraging in that aspect, though. Which was really cool to see him follow in that manner. Other than that everyone kind of took the initiative in what they wanted to do except I guess myself and maybe Student D, because we already had a clear idea in mind, especially because the idea was already suggested by me, so we took the role of more of the vision of it to make sure it was understood by everyone. But then, people were just like: Oh, I'll do the introduction! Oh, I'll do this part and some people were just like, ok, I'll take what's left, it doesn't matter to me.

Q: And how did you guys communicate?

A: Any of the little things for which we needed a quicker response we used WhatsApp in a group chat and then any exchange of major data or writing and especially when we were writing our paper together and working on the presentation, and things like that, it was all through Teams. Or people drive and then … We actually did a lot of partnering, so if you had a section, you and your partner did it together and then you added that to the group and then had someone check your work and to write. So, it would be on Google Drive or on Teams, for you and your partner and then on Teams for the final draft.

Q: That's a lot of platforms to manage, was it too much?

A: No, because people didn't really abuse them and communicate like every day or all day. It was really like: Ok, you post something either when you need information from people or when you put in your information into the drafts.

Q: And general feedback of your experience with this course …

A: I loved it. I loved the ability to be able to do it. Especially for me having a busy schedule and being able to Skype in, to be able to be anywhere and to participate, and to be able to work with my group members at any time, at times that were more convenient for me. The remote work was huge for me. I love presenting, so I liked the idea that I had to present via Skype to IKEA, as well as I got to do it in person. So, I got kind of the feel of both, because I think that's kind of what will translate into real-life in the future, you will have to do both. Working with people from different groups was great. I wish we had more people from (the assigned country of implementation) as well, because it kind of would've made our implementation easier at the original site that was given to us, but we had to switch things around. But that part was easy to be flexible with. I think it's hard having three professors, because they each have different styles, so you need to kind of make sure you work towards all of them, but not lose the authenticity of what you really want to portray. But you need to still try to make sure you get some proper infor-mation. That part for me was actually the trickiest in my mind when writing. I felt like one professor really wanted the statistical data and one was kind of in between, she saw the benefit of both and another one being like: Ok, if you defend it, no matter what you put, as long as you come up with why you did it, it doesn't matter. So, that was kind of like something that I noticed, which was on the side, or maybe it was my own thinking that that's how it was, that I made up.

Q: That's definitely translatable to the real work ☺

A: Oh yeah, so it was definitely great, but that was maybe my weakness that I was like: Oh, so now I need to make sure that I'm satisfying everyone and I'm still keeping true to myself and what my idea is and believing in it. And I think what was most of use was that I was able to work with the company and it was something that I also believe in and it made the work so much more valuable, and it was something I was 100% in. I almost would lose myself in that coursework or that material, rather than focusing on other work that needed to be done, because it was so much fun and my group was fun. When people are getting along it makes the work a lot more entertaining and productive.

M: Sure … Things for improvement for the next year?

CR: In general, I would say information exchange between professors and us, as far as the last-minute dates … But I guess it is something of my personal preference due to my busy schedule. To have the due dates right at the beginning, but I kind of liked it … I don't know actually, it's tough. Because, I kind of liked that it was like no, now it's 'x' day. Do it! I would kind of say that I liked both …

Q: I wanted to ask you before about the space in which you were working, because as you said – in virtual work, virtual teams it is hard to define space, or you don't know where people are working and the great part of it is that it can be anywhere. But some people still choose to work in the office, or they prefer to work from home (or wherever). And I've found that for some people it depends either on their personality or on the kind of work they have to do. And how was it in your case and in the case of your team? Where did everyone work and where did you connect from? And was it that some works you performed in some kind of spaces and the others in different ones?

A: I think during a lot of our meetings people where usually at home. Because a lot of them were in the evenings when classes had ended, so actually I can't recall a time when our meetings were at … where people weren't in their home space. So, then you actually got to see a little, you see the background and you kind of … not to judge, but you think: Oh … ok that's how this person's living space is like.

Interview 2

Q: What do you think influences the collaboration – personality, culture, availability, interest in project, motivation?

A: All of the above, especially motivation, dedication, personality (how they work), being organised or not (not the actual amount of time available, but its management).

Q: And the diversity, e.g. age, gender, nationalities, languages spoken, universities, backgrounds in education, ways of working – how did it affect the work?

A: If you do not have many (mentioned) things in common in a team, you need to have the same etiquette of work. E.g. if you say something, you deliver it and you deliver it on time ... I liked Student D's style of work: she was quick at replying, she did a really good job ... you may think we don't have much in common, but we got along very well in terms of the way we work.

Q: Some guidelines, but few limitations and restrictions, a lot of freedom (of space, way of working and learning) – how did it affect your teamwork?

A: It was good that we had the freedom, but what was really missing was enthusiasm, motivation and dedication, which I was expecting. If some people are engaged and others are not, it is frustrating for both sides. They (less active team members) feel pushed – they have different incentives, they just want to pass the course, not caring as much about the performance, while the others have to take over the workload to increase the performance of the entire group ...

Q: We gave you data and guidelines, but no strict expectations. Each group is unique, and we wanted to see what you would do with the information you get and how you would approach it (if it would trigger ideas, productivity, teamwork or be a reason to put less effort). Would a weight assigned to the partial grade related to the task be a possible influence on how students approach it and how much effort they put in it?

A: Yeees ... Maybe. I don't know. I wish the communication was better in the team ... I'm actually curious how the other teams managed problems like ours. Maybe some people just got along really well, and they did not have problems ... I think it may also be a problem if in a team there are too many people that would like to lead. It is maybe something that happened with us and at the end we had many disagreements. I think the leader usually has a very strong vision, a very precise vision of things and if you have too many people like that in a team it can be quite difficult to reach some of the goals.

Q: Do you think it can work to be in a team where there is no leader, which means leaders emerge based on tasks. There is no fixed structure, but the structure is dynamic like the project is. So, for example: today I'm a leader of a meeting, tomorrow you are the leader of designing the leaflets and maybe the other one is a leader of preparing the presentation and will control that things are done on time and put together. Do you think it is possible to have a team in which things would be dynamic and would evolve smoothly without actually setting who does what? Do you think it can work efficiently?

A: Yes, I think so. It can work if the team members know each other a bit better, so they know for example their strengths. So, I know for example this person is very good with the digital skills and they can design things. I do not know, maybe ... if you do not know each other it is kind of hard to jump in and say: I will do this.

Q: So, people didn't jump in and that was the problem?

A: Some people did, some did not.

Q: So the idea is maybe to get people engaged (and responsible) enough, so that they would always want to jump in and instead of being called to do things, they would be called not to do things … Do you think that could work?

A: Yes, that is how I imagine it would work!

Q: So that's the default state for you?

A: Yes, also because of the nature of the project. On one hand, there was some sort of competition, but on another not really. It is a project, that is for everyone's good and also in theory everybody involved should have been excited about it. So, I thought everybody would want to jump in and do things.

Q: Was there anything other than the motivation that was disturbing the teamwork?

A: Maybe the different schedules, because everybody had their own schedule and some people had a lot of classes and others had less and people had different engagements and it was kind of hard. Because we had to work for certain deadlines, but to make everybody align for this deadline, but maybe they had something else important then. I think if you are at work it is not the same, because you all have this as the main thing and know this is the time to work on it.

Appendices

FIGURE 14.1 Student teams getting ready to gather observations on food waste in the back kitchen of the IKEA store in Stockholm (February 2018)

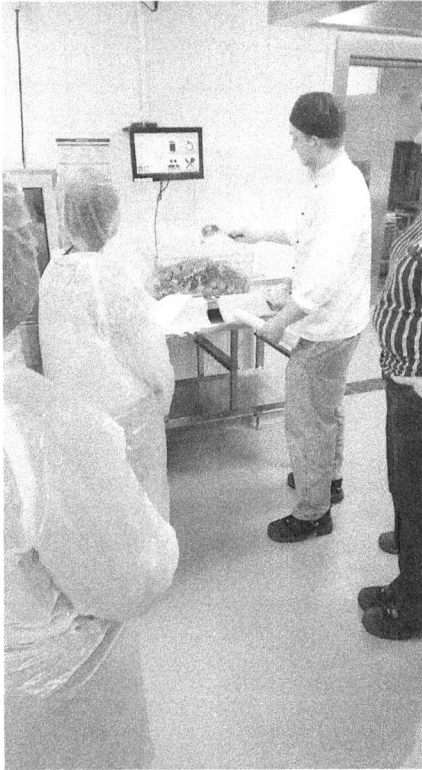

FIGURE 14.2 Students learning about the IKEA's Food Waste Watcher Programme in the Stockholm store kitchen (February 2018)

FIGURE 14.3 Kick-off meeting lecture in Stockholm School of Economics (February 2018)

FIGURE 14.4 The interactive Cross-Cultural Training at the kick-off in Stockholm School of Economics (February 2018)

FIGURE 14.5 Dinner at the kick-off in Stockholm School of Economics (February 2018)

FIGURE 14.6 Virtual meeting between three Universities – Lugano perspective (April 2018)

FIGURE 14.7 The final summit in Lugano – opening with the Rector (May 2018)

FIGURE 14.8 The final summit in Lugano – student presentations (May 2018)

FIGURE 14.9 The final summit in Lugano – group photo (May 2018)

FIGURE 14.10 The final summit in Lugano – 'Master Chef' style dinner (May 2018)

15

CREATING SHARED VALUE FOR COMPANIES IN TACKLING WORLD CHALLENGES

Tatiana Egorova and Marijane Luistro-Jonsson

Introduction

Courses promoting partnership with companies, like Tackling World Challenges, is not new. More than 30 years ago, collaboration between universities and industry occurred and proved to be desirable for companies to remain competitive (Azaroff, 1982). These partnerships involve the practice of creating 'value' (Vauterin, Linnanen & Marttila, 2012), in its multitude of definitions. For the purpose of this chapter, we define value as the benefit extracted and harnessed by some or all stakeholders. Studies suggest, for instance, that cooperation between universities and industry provides companies with new knowledge, know-how and knowledge transfer (Gattringer, Hutterer & Strehl, 2014; Siegel, Waldman, Atwater & Link, 2003). As this type of collaboration became common (Slotte & Tynjälä, 2003), the outcomes and the value produced have been subject to significant variation (Howells et al., 2012), resulting in ambiguity on what the collaboration offers. On one hand, there are thoroughly researched and established studies focusing on value creation through a research and development collaboration set-up between industry and universities (e.g. Mindruta, 2013; Emden, Calantone & Droge, 2006), mostly in the form of a supervisory thesis and internship projects. On the other hand, there are less researched areas such as investigating how value is created among thematic and methodological teaching courses involving industry participation. We address this imbalance in the literature by analysing the process of value creation for the industry partner using the lens of shared value creation within the setting of a specific course taking new pedagogical approaches.

The reason we focus in this direction is twofold. First, as the nature of the value creation that companies want to impart and leave behind shifts from a purely profit-oriented approach to one that considers sustainability aspects, companies seem to lack the necessary capabilities and expertise on how to deal with certain of these issues such as climate change. It has been found that while 90% of executives realise the importance of sustainability related matters, only a quarter of them know what actions should be taken in this respect (Kiron et al., 2017), indicating that sustainability literacy and skills are currently lacking in the industrial setting. Secondly, with the advancement of technology, new platforms have emerged for collaboration across different actors. It has been determined that the education of a new generation of employees who are comfortable and skilled in the new information society is of vital importance for the industry and academia alike (Slotte & Tynjälä, 2003). This allows collaboration not just with one educational institution but various ones at the same time, which can give richness in diversity of ideas and impact among projects. Naturally, there are obstacles in the process. Even though the benefits of such a collaboration may be recognised by top management, the actual burden of working with students often requires an investment of additional time for regular employees who as a result may not welcome such interactions (Kock, Auspitz & King, 2000). Given these factors, we attempt to revisit and analyse, in this new setting, the value that the course creates using the theoretical framework of shared value creation. We first review the shared value concept and the avenues of value creation, as well as its critiques; next we analyse the value created by the course for companies and broader society; and finally, we conclude with a discussion of the value created in direct and indirect ways.

Shared value as a theoretical framework

Value creation is a concept of great interest for scholars and practitioners alike (Lepak et al., 2007), even if its conceptualisation has been highly elusive due to various interpretations of what is value. One of the approaches particularly favoured by companies is the creation of 'shared value' (e.g. Pfitzer, Bockstette & Stamp, 2013). The idea of shared value, proposed by Kramer & Porter (2011), is to centre business activities on social needs. In essence, shared value implies that the economic profits and corporate social responsibility (CSR) should not be viewed as distinct directions of the company's activity, but as its essential components blended together. Thus, this approach shifts the focus of attention from merely increasing shareholder value to creating value for society while, at the same time, making profits, which can be done by reconceiving products and markets, redefining productivity in the value chain and enabling local cluster development (Kramer & Porter, 2011). We briefly present these strategies or avenues for shared value creation below.

Reconceiving products and markets

Viewing products and services through the lens of societal needs opens up a highway for innovation and the creation of shared value. The first part of the reconceiving process deals with products and services that should be aligned with the growing demand for more sustainable, healthier, climate friendly, less wasteful and more responsible offerings. With these issues in mind, new redesigned products not only reach a broader audience but entail greater societal impact, often in terms of public health and environmental improvements. The second part is to identify and approach new markets with suitable products and services. These efforts may include targeting other markets, for example, the so-called 'bottom of the pyramid', disadvantaged communities or other underserved markets, as well as re-conceptualising what constitutes a market. For example, the market could be built around avoiding or mitigating some harm instead of serving a particular need. Shared value thinking, through reconceiving products and markets, can give companies a competitive advantage (Kramer & Porter, 2011).

Redefining productivity in the value chain

Value chains offer numerous opportunities for shared value creation as they usually involve multiple societal problems, such as environmental degradation or unfair practices. Inefficiencies in value chains often represent costs for the company and the society, for example, inefficient transportation solutions are both costly and create environmental externalities. Therefore, improvements along the value chains may have synergistic effects in terms of shared value creation. While there are many improvement opportunities that can be pursued, the following areas deserve primary attention: energy use and logistics, resource use, procurement, distribution and employee productivity (Kramer & Porter, 2011).

Enabling local cluster development

Companies do not exist in a vacuum. The institutions and structures surrounding them are of crucial importance for shared value creation. Geographical clusters are known for facilitating innovation (Porter & Stern, 2001) and enhancing productivity (Porter, 2000), both of which are closely related to the creation of shared value as they are crucially important for the first two avenues, namely, reconceiving products and markets and redefining productivity in the value chain. Clusters are conducive for synergies in terms of product development, make procurements more efficient and bring about such spill over effects as the creation of new businesses and attraction of talent (Kramer & Porter, 2011).

Shared value critique

In sum, the idea of shared value simply means that value for the company and society can be co-created through one or a combination of the strategies described above. This sweeping statement has been subject to criticism from several directions. Many share the view that the concept is not something new and overlaps with other concepts in CSR and social innovation literature, which was never duly acknowledged by its authors (Crane et al., 2014). The concept is vaguely defined, creating complications in operationalising and measuring shared value (Dembek,

Singh & Bhakoo, 2016), and is mostly based on anecdotal evidence. It is also argued that shared value as a concept is unfairly contrasted with the traditional understanding of CSR, which is best characterised as philanthropy, while the modern view on CSR is superior to the creation of shared value in many aspects (Beschorner & Hajduk, 2017). In addition, shared value has been blamed for not being radical enough to spur real change (Beschorner, 2014).

Despite the critical voices, shared value is a simple framework and popular approach that companies often find practical and feasible to use (e.g. Pfitzer, Bockstette & Stamp, 2013). Multinational companies such as Adidas, BMW, Heinz (Klein, 2011) and Novartis (Kramer, 2012) are engaging in the creation of shared value. The growing evidence that companies have begun to innovate their business models with the purpose of creating shared value (Michelini & Fiorentino, 2012) indicates that business society finds shared value applicable and possible to integrate into their business activities. Using this framework, which speaks to practitioners, we will discuss in the next section how the course can create shared value for and together with the companies.

How the course creates shared value for the companies

The TWC course, as described in the other chapters, can be characterised by the following dimensions: internationalisation, digitalisation, practical orientation and focus on sustainability. We analyse in the ensuing paragraphs how each of these dimensions contributes to the creation of shared value, focusing on the corporate partners.

All projects developed by participating students have two points of departure: the societal problem of food waste and the practical issue faced by the partner company in this respect. Thus, by design, students are bound to develop a solution that will create value for the company and, at the same time, for the greater society. Other components of the course, namely, digitalisation and internationalisation, contribute to the creation of shared value as well. While digitalisation acts as an enabler of communication and development of the project, the internationalisation component brings about the diversity of perspectives that allow students to see products and markets from different angles, assess value chains in a more holistic way and essentially develop a network that shapes the cluster of companies and universities in the European Union.

The projects developed by students in the course follow one or several avenues of shared value creation. We discuss two examples of resulting solutions that focus on the same problem of reducing food waste of particular vegetable products of the manufacturer. The company was relatively satisfied with their food waste management; however, one of the student groups managed to identify several potential directions of value creation by applying The Waste and Resources Action Programme (WRAP) methodology (WRAP, 2016). In particular, they proposed ways to reduce food waste during tomato picking and transportation stages, essentially improving the productivity along the value chain, and engaging in coopetition at the industry level by creating a local cluster. Another student team working on the same issue took a broader look at the value chains and proposed ways to engage consumers in the value creation process via communication on the packaging. These examples illustrate how student ideas and recommendations on redefining productivity in the value chain and enabling local cluster development can create shared value for the industry partner. While reducing food waste, these efforts have the potential of lowering economic costs and building consumer loyalty for the company as well, which results in value creation for the wider society as well as the industrial partner.

Students also often follow the third avenue of shared value creation and reconceive products and markets. For example, in dealing with the same issue, another group of students focused on ways to create shared value out of coffee silverskins – a common waste produced in the coffee roasting process. Their idea was to use silverskins as raw material for some products, and initially they considered applications in the cosmetics industry and ink production. However, they came to the conclusion that the former will result in even greater levels of pollution than the generation of waste in the form of silverskins, while the latter would be technologically unfeasible because of the low pigment content in silverskins. Eventually, the students proposed using coffee silverskins as the input for the production of biogas, which can be used both for the internal needs of the roasting company and sold to external consumers. Essentially, the team developed a new way to view waste as a product (biogas) that opens a door to a new market (fuel and/or energy supply). In addition, during the process of reconceiving waste as a product, the students critically assessed other potential applications from the point of view of shared value creation and decided against them.

Similarly, in working with the restaurants of retail stores and school cafeterias, the student groups often tried to influence the behaviour of consumers to minimise food waste on the plate, through simple behavioural nudges by introducing new information, reminders, mindfulness techniques, as well as interactive activities with children or even the idea of a food waste museum. New financial products were also suggested as they analysed contracts among municipalities and foodservice providers. This variety of approaches to influence market behaviour in minimising food waste similarly reflects reconceiving current markets as they aim to transform current market behaviour, as well as creating a new ecosystem and clusters of support for behavioural change. This process of creating shared value in the projects trains the students to develop an important skill that employers would want their future employees to have and allows companies to screen potential future employees among the students.

Despite the great potential of shared value creation, the structure of the course has generally some drawbacks that can hinder the process. The internationalisation component is inevitably linked to a higher instance of intercultural conflicts that may impede the creation of ideas among the students. In turn, while facilitating and enabling the course flow, digitalisation leads to the lack of direct and personal contact among the students as well as between the students, the companies and teaching team, which may hamper trust and communication (Jarvenpaa & Leidner, 1999). However, since a central goal of the course is to expose students to working together across the different educational institutions digitally, it is very important and beneficial for their future careers to acquire skills of working in virtual and international teams early, such as intercultural skills.

Moreover, when it comes to practical orientation, it is hard to find companies that are willing to work with students. This lack of motivation can limit the number of potential projects that students could work on. Finally, the focus on sustainability and particularly on the issue of food waste constrains the attention of students and may be suboptimal for the companies that could possibly benefit more from directing attention towards other sustainability issues that may be more pressing in their case. Figure 15.1 gives an overview of how each component of the course can strategically create shared value, as well as its possible drawbacks.

	Internationalisation	*Digitalisation*	*Practical orientation*	*Focus on sustainability*
Reconceiving products and markets	Diversity of views on products and markets	Enabler of the whole process that brings together all parts of the course	Solutions are developed for the companies based on their needs, with close collaboration with the representatives of the company and the guidance of the teaching team. Essentially, they constitute full scale consulting projects.	Projects depart from a sustainability issue of food waste with the clear goal of reducing it
Redefining productivity in the value chain	Value chains are viewed from the global perspective			
Enabling local cluster development	Cluster of companies and universities in the EU			
Negative implications	Potential intercultural conflicts may hinder the creation of shared value	Lack of direct contact may lead to misunderstandings and lower efficiency	Lack of interest on behalf of the companies	Narrow focus of one problem hinders potential improvements in other sustainability issues faced by the company

FIGURE 15.1 Shared value creation strategies by each component of the course

Final remarks

The analysis presented in the previous section shows some support that the course creates value for the companies and the broader society in direct and indirect ways. The direct way is the development of the projects themselves. While the projects are characterised by great variety, they generally produce value along the lines of the shared value framework, i.e. by reconceiving products and markets, redefining productivity in the value chain and enabling local cluster development. Indirectly, the other components of the course, such as internationalisation and digitalisation, contribute to the creation of shared value that can eventually be harnessed by the partner companies. Essentially, the combination of these two approaches can be a possible avenue for building the business case for sustainability in business education.

Despite the critical voices suggesting that the concept of the 'business case for sustainability' may not be enough to cope with the magnitude of the problem (Landrum & Ohsowski, 2018), companies indeed continue to work in this direction and would like their future employees to possess relevant skills (Kiron et al., 2017). In this respect, it is important to keep in mind that the course is primarily an educational endeavour with the goal of educating students and preparing them for their working life. This constitutes the indirect way creating value for the business and society – by training skilled future employees of the companies. The course alumni are comfortable with new technology, have skills for working in virtual teams that were identified as crucial for the industry and academia (Slotte & Tynjälä, 2003) and have gained experience in practically tackling sustainability issues, which is a rare and valuable skill (Kiron et al., 2017). Thus, the companies benefit from the course immediately during the project development phase and over a longer term by enjoying access to a highly and relevantly skilled workforce.

Our experience suggests that companies should be more open to such collaborations as eventually they may enjoy numerous benefits such as new ideas, projects that are ready to be developed and bring value, access to a talent pool of highly skilled potential employees as well as branding and image building opportunities. In this respect, communication of the value created by the collaboration between the business community and universities, not only within the R&D setting but also in the classrooms, is of vital importance.

The analytical framework presented in this chapter is not constrained by the boundaries of this course and can be employed for assessment of other courses, modules or projects that involve collaboration between university students and external parties. The proposed approach to analysing these endeavours provides a clear view of the benefits and disadvantages in terms of value creation for the industry partner. The framework further offers some flexibility to assess value creation not only from the standpoint of companies participating in the course but also from the point of view of other stakeholders. Finally, the lines of analysis are not fixed and can be replaced based on specific requirements. Thus, one can argue that the proposed framework is a useful analytical tool that can be used by researchers and practitioners alike.

References

Azaroff, L.V. (1982). Industry–university collaboration: How to make it work. *Research Management*, 25(3), 31–34.

Beschorner, T. (2014). Creating shared value: The one-trick pony approach. *Business Ethics Journal Review*, 1(17), 106–112.

Beschorner, T. & Hajduk, T. (2017). Responsible practices are culturally embedded: Theoretical considerations on industry-specific corporate social responsibility. *J Bus Ethics*, 143, 635–642. Retrieved from https://doi.org/10.1007/s10551-016-3405-2

Crane, A., Palazzo, G., Spence, L.J. & Matten, D. (2014). Contesting the value of 'creating shared value'. *California Management Review*, 56(2), 130–153.

Dembek, K., Singh, P. & Bhakoo, V. (2016). Literature review of shared value: A theoretical concept or a management buzzword? *Journal of Business Ethics*, 137(2), 231–267. doi:10.1007/s10551-015-2554-z

Emden, Z., Calantone, R.J. & Droge, C. (2006). Collaborating for new product development: Selecting the partner with maximum potential to create value. *Journal of Product Innovation Management*, 23(4), 330–341

Gattringer, R., Hutterer, P. & Strehl, F. (2014). Network-structured university-industry-collaboration: Values for the stakeholders. *European Journal of Innovation Management*, 17(3), 272–291.

Howells, J., Ramlogan, R. & Cheng, S.L. (2012). Innovation and university collaboration: Paradox and complexity within the knowledge economy. *Cambridge Journal of Economics*, 36(3), 703–721.

Jarvenpaa, S.L. & Leidner, D.E. (1999). Communication and trust in global virtual teams. *Organization Science*, 10(6), 791–815.

Kiron, D., Unruh, G., Kruschwitz, N., Reeves, M., Rubel, H. & Felde, A. (2017). Corporate sustainability at a crossroads: Progress toward our common future in uncertain times. Retrieved from http://marketing.mitsmr.com/offers/SU2017/58480-MITSMR-BCG-Sustainability-Report-2017.pdf

Klein, P. (2011). Three great examples of shared value in action. Retrieved from www.forbes.com/sites/csr/2011/06/14/three-great-examples-of-shared-value-in-action/

Kock, N., Auspitz, C. & King, B. (2000). Using the web to enable industry-university collaboration: an action research study of a course partnership. *Informing Science*, 3(3), 157–165.

Kramer, M. (2012, 25 April). Better ways of doing business: Creating shared value. *The Guardian*. Retrieved from www.theguardian.com/sustainable-business/blog/creating-shared-value-social-progress-profit

Kramer, M.R. & Porter, M. (2011). Creating shared value. *Harvard Business Review*, 89(1/2), 62–77.

Landrum, N.E. & Ohsowski, B. (2018). Identifying worldviews on corporate sustainability: A content analysis of corporate sustainability reports. *Business Strategy and the Environment*, 27(1), 128–151.

Lepak, D.P., Smith, K.G. & Taylor, M.S. (2007). Introduction to special topic forum: Value creation and value capture: A multilevel perspective. *The Academy of Management Review*, 32(1), 180–194. Retrieved from https://doi.org/10.2307/20159287

Michelini, L. & Fiorentino, D. (2012). New business models for creating shared value. *Social Responsibility Journal*, 8(4), 561–577. Retrieved from https://doi.org/10.1108/17471111211272129

Mindruta, D. (2013). Value creation in university-firm research collaborations: A matching approach. *Strategic Management Journal*, 34(6), 644–665.

Pfitzer, M., Bockstette, V. & Stamp, M. (2013). Innovating for shared value. *Harvard Business Review*, 91(9), 100–107.

Porter, M.E. & Stern, S. (2001). Innovation: Location matters. *MIT Sloan Management Review*, 42(4), 28–36.

Porter, M.E. (2000). Location, competition, and economic development: Local clusters in a global economy. doi:10.1177/089124240001400105

Siegel, D.S., Waldman, D.A., Atwater, L.E. & Link, A.N. (2003). Commercial knowledge transfers from universities to firms: Improving the effectiveness of university–industry collaboration. *The Journal of High Technology Management Research*, 14(1), 111–133.

Slotte, V. & Tynjälä, P. (2003). Industry–university collaboration for continuing professional development. *Journal of Education and Work*, 16(4), 445–464.

Vauterin, J. J., Linnanen, L. & Marttila, E. (2012). Value creation in international higher education: The role of boundary spanning in university-industry collaboration. *International Journal of Quality and Service Sciences*, 4(3), 283–298.

WRAP (The Waste and Resources Action Programme). (2016) Quantification of food surplus and waste. Retrieved from www.refreshcoe.eu/wp-content/uploads/2017/06/WRAPQuantification-of-food-surplus-and-waste-May-2016-Final-Report-Summary.pdf

16

WICKED LEARNING

An evolving recipe

Marijane Luistro-Jonsson, Carol Switzer, Michael Gibbert and Liisa Välikangas

Drawing by Inês Costa

Checklist of ingredients

The ingredients exemplify the strategic aims of the project and can vary a great deal. In developing the TWC course, we started by thinking that the old system is broken (in particular, the expulsion of Switzerland from Erasmus which created space for innovative ideas on internationalisation). If we are starting from scratch, what would we wish for? Hence, we came up with key ingredients of the course as a blender for achieving diversity, enhancing learning and rethinking international encounters.

Are the following elements in place?

- Alignment of respective goals of participating educational institutions
- Commitment of industry and other international partners

- IT involvement and support
- Updated toolkit with relevant methods
- Flexibility of faculty members to go beyond disciplines
- Initial face-to-face encounter and group interactions among students
- Culminating face-to-face interactions (optional)
- Learning pedagogical tools are blended and flexible
- Diversity among participants (gender, background, culture, discipline)

Which among the ingredients above can you use in your course? What would you like to add?

_____

_____

_____

_____

_____

_____

_____

_____

_____

_____

Method and directions

Cooking involves developing dishes with a dash of creativity and experimentation when combining ingredients. In this course, we add collaboration to the mix, which can spice up otherwise static methods. Importantly, we do not advocate a single winning procedure for how to achieve a successful mixture and combination of ingredients; instead, we identify the following crucial challenges in the process:

1. *Establish the programme as a collaboration between different actors.* This entails having the setting in place: commitment of resources among the different schools, industry partners and faculty; agreed design and division of responsibilities; and effective communication of potential contributions of the projects to stimulate partner and student engagement.

2. *Preparation and Pre-launch.* This would involve gathering and preparing key ingredients: coordinating the schedule of faculty and experts, promoting the course to prospective students, making arrangements for the kick-off event and travel plans, and finalising course details, projects, rules and requirements. This is to ensure that group dynamics skills and support are in place to enhance the potential synergy among the students, and that course content can effectively impart sustainability literacy and a fitting methodological toolkit for the projects.

3. *Course implementation.* The implementation would require fulfilling expected roles and commitments, guiding the students in realising their ideas and providing them with appropriate steps and methods, as well as improvisation when problems occur. Efficient communication among the different actors, particularly in digital platforms, is crucial to ensure the smooth flow of the course.

4. *Feedback.* This part is twofold. Firstly, create space to enjoy the product of a semester's work, as you would when sitting down to share a fine dinner with your team. A warm ambience is conducive to reflection and the entrepreneurial spirit that encourages innovation. Then, jump in! The second part refers to processing learnings, updating knowledge and improving the course recipe in preparation for the following season.

In this final chapter, we invite you to be part of the feedback process – particularly to build upon our lessons learned and further experiment to generate new ideas and ultimately come up with a better recipe. The cycle continues entailing an iterative continuous improvement of the learning recipe.

Which steps are important and relevant for your course? Which ones are new? Aside from those indicated above, what additional steps would you like to add or how would you tweak this recipe for your own needs?

_____

_____

_____

_____

_____

_____

_____

Food for thought

In the process of being part of an evolving pedagogical recipe that has enabled students to address the food waste issue for the past four years, we have come up with some considerations that can guide educators who are on the same journey. In summary, our general lessons learned cover (1) the need for bold intent and togetherness, (2) modest but creative experimentation, and (3) reflexivity in a pre-cise (challenge or problem) context. In particular, we have gained the following specific nuggets of understanding:

The 'art of keeping going' even when students are faced with a wicked problem on the scale of a grand challenge, or as the faculty face setbacks related to the course. As we encourage the students to act and experiment with the partners in addressing difficult problems, it exposes them to the sometimes bleak realities and inhibiting complicities of life, imparting the insight that they can make modest differences, yet even these drops in the ocean count.

The students are in charge of the projects and are fuelled with a sense of responsibility, aiding them to realise the value of group effort and team dynamics. In the same manner, as faculty members who direct and coordinate the course, we realise the importance of inclusiveness as we work with IT, administration, students, experts and other faculty members to maintain the flow. Tempering bold intent with the realisation of the need for togetherness has thus been a lesson learned not only by the students in actualising their project ideas but for the faculty as well in actualising this course.

Possible solution ideas are modest but may have global scale due to the reach of a corporate or institutional partner. Collaborating with companies can be quite tricky due to conflicting agendas and any unexpected turn of events; however, the importance of these collaborations should not be overlooked. Effective communication of the potential contributions of student projects can increase the engagement of all actors involved.

Virtual and digital contact are not sufficient alone. One should not underestimate the importance of face-to-face interactions in aiding trust-building and sustained cooperation among group members who are geographically dispersed, culturally diverse and of different disciplinary backgrounds.

Science and specific disciplinary knowledge are needed. In order to impart sustainability literacy, the role of science comes into the picture to give systematic structure and evidence-based support to address wicked problems. The toolkit we provide aims to augment and create synergy with the multidisciplinary backgrounds and rich empirical experiences the students bring with them.

Students need to care for the issue enough to persist in the work. We also note that inspiration and engagement is important to sustain the ambiguity and uncertainty that the students may face in their journeys.

When things do not go as expected, here are additional important things to hasten resilience:

- Ask for collegial support and help in the case of institutional problems such as lack of funding
- Simplify communication channels for swift decision-making
- Present new technological platforms but let the students decide what works for them
- Informal settings for group dynamics work better than forced ones
- Avoid free-riding by designing projects requiring efforts from each team member.
- Inoculate students to manageable levels of ambiguity for them to develop their creativity and flexibility, but not so much that it discourages their enthusiasm and engagement.

Finally, one should keep in mind that courses of this nature require organising outside organisations. Thus, in partially organising, we need to be mindful of balancing dominance of contrasting logics and the representation among different actors involved. Learning the art of giving space to all those involved is at the heart of collaborating with others.

What other food for thought and insights come to mind? Are there other lessons that you can add in improvising solutions for unforeseeable situations?

A wicked challenge

Finally, we would like to end this book by challenging you to try and improve on our Wicked Learning recipe. We hope that you join us and continue the mission of Tackling World Challenges and sustainability issues in our increasingly internationalised and digitalised world. Remember to have fun at every step of your journey – from choosing and designing your recipes, setting up and cooking the course, feeding ideas and chewing on the lessons learned, and sharing them with the wider community. As you join us in this process by completing the lines with your own novel ideas and innovative pedagogical measures, may you further develop and experiment on your own wicked pedagogical recipes that tackle world challenges. We invite you to take on this wicked challenge.

Things to do for my wicked challenge:

_____

_____

_____

_____

_____

_____

_____

_____

_____

_____

_____

For Product Safety Concerns and Information please contact our EU
representative GPSR@taylorandfrancis.com
Taylor & Francis Verlag GmbH, Kaufingerstraße 24, 80331 München, Germany

www.ingramcontent.com/pod-product-compliance
Lightning Source LLC
Chambersburg PA
CBHW070717220326
41598CB00024BA/3195

* 9 7 8 0 3 6 7 1 9 7 6 2 9 *